HENRY MARSH studied medicine at the Royal Free
Hospital in London, became a Fellow of the Royal
College of Surgeons in 1984 and was appointed
Consultant Neurosurgeon at Atkinson Morley's/
St George's Hospital in London in 1987. He has
been the subject of two major documentary films,
Your Life in Their Hands, which won the Royal
Television Society Gold Medal, and *The English
Surgeon*, which won an Emmy. He is married to
the anthropologist and writer Kate Fox.

Accolades for *Do No Harm*

Winner of the PEN Ackerley Prize

Winner of the South Bank Sky Arts Award for Literature

Shortlisted for the Guardian First Book Award

Shortlisted for the Costa Book Award

Shortlisted for the Duff Cooper Prize

Shortlisted for the Wellcome Book Prize

Shortlisted for the Slightly Foxed Best First Biography Prize

Longlisted for the Samuel Johnson Prize for Non-Fiction

Named a Best Book of the Year by *The Economist*,
Financial Times, and *Kirkus Reviews*

Additional Praise for *Do No Harm*

"Neurosurgery has met its Boswell in Henry Marsh. Painfully honest about the mistakes that can 'wreck' a brain, exquisitely attuned to the tense and transient bond between doctor and patient and hilariously impatient of hospital management, Marsh draws us deep into medicine's most difficult art and lifts our spirits. It's a superb achievement." —Ian McEwan

"When a book opens like this: 'I often have to cut into the brain and it is something I hate doing'—you can't let it go, you have to read on, don't you? Brain surgery, that's the most remote thing for me, I don't know anything about it, and as it is with everything I'm ignorant of, I trust completely the skills of those who practice it, and tend to forget the human element, which is failures, misunderstandings, mistakes, luck and bad luck, but also the non-professional, everyday life that they have. *Do No Harm: Stories of Life, Death, and Brain Surgery* by Henry Marsh reveals all of this, in the midst of life-threatening situations, and that's one reason to read it; true honesty in an unexpected place. But there are plenty of others—for instance, the mechanical, material side of being, that we also are wire and strings that can be fixed, not unlike cars and washing machines, really."
—Karl Ove Knausgaard, *Financial Times* (London)

"Marsh, one of our leading neurosurgeons, is an eloquent and poetic writer. *Do No Harm* offers a rare behind-the-scenes look at the most mysterious part of human life. His descriptions of neurosurgery are at once fascinating and illuminating; a gripping memoir of an extraordinary career."
—Daniel J. Levitin, PhD, author of
The Organized Mind and *This Is Your Brain on Music*

"The outstanding feature of *Do No Harm* is the author's completely candid description of the highs and lows of a neurosurgical career. . . . For its unusual and admirable candor, wisdom and humor, *Do No Harm* is a smashing good read from which the most experienced and the most junior neurosurgeons have much to learn."   —*AANS Neurosurgeon*

"One of the best books ever about a life in medicine, *Do No Harm* boldly and gracefully exposes the vulnerability and painful privilege of being a physician."

—*Booklist* (starred review)

"This thoughtful doctor provides a highly personal and fascinating look inside the elite world of neurosurgery, appraising both its amazing successes as well as its sobering failures."   —*Publishers Weekly* (starred review)

"His love for brain surgery and his patients shines through, but the specialty—shrouded in secrecy and mystique when he entered it—has now firmly had the rug pulled out from under it. We should thank Henry Marsh for that."

—*The Times* (London)

DO NO HARM

DO NO HARM

Stories of Life, Death,
and Brain Surgery

Henry Marsh

PICADOR

A Thomas Dunne Book
St. Martin's Press
New York

picadorusa.com • picadorbookroom.tumblr.com
twitter.com/picadorusa • facebook.com/picadorusa

Picador® is a U.S. registered trademark and is used by St. Martin's Press under license from Pan Books Limited.

For book club information, please visit facebook.com/picadorbookclub or e-mail marketing@picadorusa.com.

An earlier version of chapter "Pineocytoma" originally appeared on granta.com.

The Library of Congress has cataloged the Thomas Dunne Books edition as follows:

Marsh, Henry, 1950–
    Do no harm : stories of life, death, and brain surgery / Henry Marsh. — First U.S. edition.
        p. cm.
    ISBN 978-1-250-06581-0 (hardcover)
    ISBN 978-1-4668-7280-6 (e-book)
    1. Marsh, Henry, 1950– 2. Neurosurgery—England—Anecdotes.
3. Neurosurgery—England—Personal Narratives. 4. Neurosurgical Procedures—England—Anecdotes. 5. Neurosurgical Procedures—England—Personal Narratives. 6. Physician-Patient Relations—England—Anecdotes. 7. Physician-Patient Relations—England—Personal Narratives. I. Title.
    RD592.8
    617.4'8092—dc23   2015002573

Picador Paperback ISBN 978-1-250-09013-3

Our books may be purchased in bulk for promotional, educational, or business use. Please contact your local bookseller or the Macmillan Corporate and Premium Sales Department at 1-800-221-7945, extension 5442, or by e-mail at MacmillanSpecialMarkets@macmillan.com.

Originally published in the United Kingdom by Weidenfeld & Nicolson, an imprint of the Orion Publishing Group Ltd

First published in the United States by Thomas Dunne Books, an imprint of St. Martin's Press

First Picador Edition: June 2016

10 9 8 7 6 5 4 3 2 1

For Kate, without whom this book
would never have been written

'First, do no harm ...'
Commonly attributed to Hippokrates of Kos, *c.* 460 BC

'Every surgeon carries within himself a small cemetery, where from time to time he goes to pray—a place of bitterness and regret, where he must look for an explanation for his failures.'
René Leriche, *La philosophie de la chirurgie*, 1951

CONTENTS

CONTENTS

PREFACE TO THE 2016
PAPERBACK EDITION

If we are ill and in hospital, fearing for our life, awaiting terrifying surgery, we have to trust the doctors treating us – at least, life is very difficult if we don't. It is not surprising that we invest doctors with superhuman qualities as a way of overcoming our fears. If the operation succeeds, the surgeon is a hero, but if it fails, he is a villain.

The reality, of course, is entirely different. Doctors are human, just like the rest of us. Much of what happens in hospitals is a matter of luck, both good and bad; success and failure are often out of the doctor's control. Knowing when not to operate is just as important as knowing how to operate, and it is a more difficult skill to acquire.

A brain surgeon's life is never boring and can be profoundly rewarding, but it comes at a price. You will inevitably make mistakes, and you must learn to live with the occasionally awful consequences. You must learn to be objective about what you see, and yet not lose your humanity in the process. The stories in this book are about my attempts, and occasional failures, to find a balance between the necessary detachment and compassion that a surgical career requires, a balance between hope and realism. I do not want to undermine public confidence in brain surgeons or, for that matter, the medical profession, but I hope that my book will help people understand the difficulties – so often of a human rather than technical nature – that doctors face.

PREFACE

Throughout my career I have been fortunate to work with colleagues from America – mainly neurosurgical residents who come to work for one year in my department in London as part of their training. I have learned much from them and as with many Brits who have worked with Americans I love their optimism, their faith that any problem can be solved if enough hard work and money is thrown at it, and the way in which success is admired and respected and not a cause for jealousy. This is a refreshing contrast to the weary and knowing skepticism of the English. Yet when I visit American hospitals and see the extremes to which treatment can sometimes be pushed, I wonder whether the doctors and patients there have yet to understand that the famous dictum that in America death is optional, was meant as a joke.

I have also worked in countries such as Ukraine and Sudan that have very impoverished health-care systems compared to America. You realize quite quickly, however, that despite the very great differences in equipment and technology many things are the same. Our vulnerability and fear of death when we are patients know no national boundaries, and the need for honesty and kindness from doctors – and the difficulty at times in giving these – is equally universal. I would hope that my many American trainees have come to understand this by working in the foreign country that is England, just as I have done with my work abroad.

Doctors will sometimes admit their mistakes and "complications" to each other but are reluctant to do so in public, especially in countries that have commercial, competitive health-care systems. This book is as much about failure as success, but it is not intended as a confession and instead is an attempt to give an honest account of what it is like to be a neurosurgeon. My readiness to admit to my fallibility is perhaps rather English, but I hope that the problems I describe will be familiar to doctors and patients everywhere. The book is also the story of an all-encompassing love affair, and an explanation of why it is such a privilege – although a very painful one – to be a neurosurgeon.

—Henry Marsh, August 2014

DO NO HARM

PINEOCYTOMA

n. an uncommon, slow-growing tumour of the pineal gland.

I often have to cut into the brain and it is something I hate doing. With a pair of diathermy forceps I coagulate the beautiful and intricate red blood vessels that lie on the brain's shining surface. I cut into it with a small scalpel and make a hole through which I push with a fine sucker – as the brain has the consistency of jelly a sucker is the brain surgeon's principal tool. I look down my operating microscope, feeling my way downwards through the soft white substance of the brain, searching for the tumour. The idea that my sucker is moving through thought itself, through emotion and reason, that memories, dreams and reflections should consist of jelly, is simply too strange to understand. All I can see in front of me is matter. Yet I know that if I stray into the wrong area, into what neurosurgeons call eloquent brain, I will be faced by a damaged and disabled patient when I go round to the Recovery Ward after the operation to see what I have achieved.

Brain surgery is dangerous, and modern technology has only reduced the risk to a certain extent. I can use a form of GPS for brain surgery called Computer Navigation where, like satellites orbiting the Earth, infra-red cameras face the patient's head. The cameras can 'see' the instruments in my hands which have little reflecting balls attached to them. A computer connected to the cameras then shows me the position of my instruments in my patient's brain on a scan done shortly before the operation. I can operate with the patient awake under local anaesthetic so that I can identify the eloquent areas of the brain by stimulating the brain with an electrode. The patient is given simple tasks to perform by my anaesthetist so that we can see if I am causing any damage as the operation proceeds. If I am operating on the spinal cord – which is even more vulnerable than the brain – I can use a method of electrical stimulation known as evoked potentials to warn me if I am about to cause paralysis.

Despite all this technology neurosurgery is still dangerous, skill and experience are still required as my instruments sink into the brain or spinal cord, and I must know when to stop. Often it is better to leave the patient's disease to run its natural course and not to operate at all. And then there is luck, both good luck and bad luck, and as I become more and more experienced it seems that luck becomes ever more important.

I had a patient to operate on with a tumour of the pineal gland. In the seventeenth century the dualist philosopher Descartes, who argued that mind and brain are entirely separate entities, placed the human soul in the pineal gland. It was here, he said, that the material brain in some magical and mysterious way communicates with the mind and with the immaterial soul. I don't know what he would have said if he could have seen my patients looking at their own brains on a video monitor, as some of them do when I operate under local anaesthetic.

Pineal tumours are very rare. They can be benign and they can be malignant. The benign ones do not necessarily need treatment. The malignant ones can be treated with radiotherapy and chemotherapy but can still prove fatal. In the past they were considered to be inoperable but with modern, microscopic neurosurgery this is no longer the case. It is usually now thought necessary to operate at least to obtain a biopsy and confirm the type of tumour so that you can decide how best to treat the patient. The pineal is buried deep in the middle of the brain so the operation is, as surgeons say, a challenge. Neurosurgeons look at brain scans showing pineal tumours with both fear and excitement, like mountaineers looking up at a great peak that they hope to climb.

This particular patient had found it very hard to accept that he had a life-threatening illness and that his life was now out of his control. He was a high-powered company director. He had thought that the headaches which had started to wake him at night were caused by the stress of having had to sack so many of his employees in the aftermath of the financial crash of 2008. It turned out that he had a pineal tumour and acute hydrocephalus. The tumour was obstructing the normal circulation of cerebro-spinal fluid around his brain and the trapped fluid was increasing the pressure in his head. Without treatment he would go blind and die within a matter of weeks.

I had had many anxious conversations with him over the days before the operation. I explained that the risks of surgery, which included death or a major stroke, were ultimately less than the risks of not operating. He laboriously typed everything I said into his smartphone, as though typing down the long words – obstructive hydrocephalus, endoscopic ventriculostomy, pineocytoma, pineoblastoma – would somehow put him back in charge and save him. His anxiety, combined with my feeling of profound failure about an operation I had carried out a week earlier, meant that I faced the prospect of operating upon him with dread.

I had seen him the night before the operation. When I talk to my patients the night before surgery I try not to dwell on the risks of the operation ahead, which I will already have discussed in detail at an earlier meeting. I try to reassure them and lessen their fear, although this means that instead I make myself more anxious. It is easier to carry out difficult operations if you have told the patient beforehand that the operation is terribly dangerous and quite likely to go wrong – I will perhaps then feel a little less painfully responsible if it does.

His wife was sitting beside him looking quite sick with fear.

'This is a straightforward operation,' I reassured them, with false optimism.

'But the tumour could be cancerous, couldn't it?' she asked.

A little reluctantly I said that it might be. I explained that I would get a frozen section during the operation – a specimen to be examined immediately by a pathologist. If he reported that the tumour was not cancerous I would not have to try to get every last little bit of tumour out. And if it was a tumour called a germinoma I wouldn't have to remove it at all and her husband could be treated – and probably cured – with radiotherapy.

'So if it's not cancer and not a germinoma then the operation is safe,' she said, but her voice tailed off uncertainly.

I hesitated, not wanting to frighten her. I chose my words carefully. 'Yes – it makes it a lot less dangerous if I don't try to take it all out.'

We talked for a little longer before I wished them good night and went home.

Early the next morning I lay in bed thinking about the young woman I had operated on the previous week. She had had a tumour in her spinal cord, between the sixth and seventh cervical vertebrae, and – although I do not know why, since the operation had seemed to proceed uneventfully – she awoke from the operation paralysed down the right side of her body. I had probably tried to take out too much of the tumour. I must have been too sure of myself. I had been insufficiently fearful. I longed for this next operation, the operation on the pineal tumour, to go well – for there to be a happy ending, for everybody to live happily ever after, so that I could feel at peace with myself once again.

But I knew that however bitter my regret, and however well the pineal operation went, nothing I could do would undo the damage that I had done to the young woman. Any unhappiness on my part was nothing compared to what she and her family were going through. There was no reason for this next operation on the pineal tumour to go well just because I hoped so desperately that it would, or because the previous operation had gone so badly. The outcome of the pineal operation – whether the tumour was malignant or not, whether I could remove the tumour or whether it was hopelessly stuck to the brain and everything went horribly wrong – was largely outside my control. I also knew that as time went by the grief I felt at what I had done to the young woman would fade. The memory of her lying in her hospital bed, with a paralysed arm and leg, would become a scar rather than a painful wound. She would be added to the list of my disasters – another headstone in that cemetery which the French surgeon Leriche once said all surgeons carry within themselves.

As soon as an operation begins, I usually find that any such morbid fear disappears. I take up the scalpel – no longer from the scrub nurse's hand but, in accordance with some Health and Safety protocol, from a metal dish – and, full of surgical self-confidence, press it precisely through the patient's scalp. As the blood rises from the wound the thrill of the chase takes over and I feel in control of what is happening. At least, that is what usually happens. On this occasion the disastrous operation of the preceding week meant that I came to the theatre suffering from severe stage fright. Instead of chatting as I usually do with the scrub nurse and Mike, one of the trainee surgeons known as specialist registrars who was assisting me, I cleaned the patient's skin and positioned the drapes in silence.

Mike had been working with me for some months and we knew each other well. I must have trained many registrars over the thirty years of my career and with most of them, I would like to think, I have got on well. I am there to train them, and must take responsibility for what they do, but they in turn are there to assist and support me and, when necessary, encourage me. I know well enough that they will usually only tell me what they think I want to hear, but it can be a very close relationship – a little, perhaps, like that between soldiers in battle – and it is what I will miss most when I retire.

'What's up, Boss?' Mike asked.

I grunted through my face mask.

'The idea that neurosurgery is some kind of calm and rational appliance of science,' I said, 'is such utter crap. At least it is for me. That bloody operation last week makes me feel as nervous as I was thirty years ago and not at all as though I was approaching retirement.'

'Can't wait,' said Mike – a standard joke the bolder of my registrars will make now that I am reaching the end of my career. There are currently more trainees than there are consultant jobs and my trainees all worry about their future. 'Anyway, she'll probably get better,' he added 'It's early days.'

'I doubt it.'

'But you never know for certain ...'

'Well, I suppose that's true.'

We were standing behind the patient as we talked, since the unconscious, anaesthetized man was propped upright in the sitting position. Mike had already shaved a narrow strip of hair away from the back of his neck.

'Knife,' I said to Agnes the scrub nurse. I took it from the dish she held out to me and quickly cut down through the back of the man's head. Mike used a sucker to clear the blood away and I then split the neck muscles apart so that we could start drilling through the bone of his skull.

'Really cool,' said Mike.

The man's scalp incised, the muscles retracted, a craniectomy of the skull performed, the meninges opened and reflected – surgery has its own ancient descriptive language – I had the operating microscope brought in and I settled down in the operating chair. With a pineal operation, unlike other brain tumours, you do not need to cut through the brain to reach the tumour; instead, once you have opened the meninges, the membrane beneath the skull that covers the brain and spinal cord, you are looking along a narrow crevice that separates the upper part of the brain, the cerebral hemispheres, from the lower part – the brainstem and cerebellum. You feel as though you are crawling along a long tunnel. At about three inches' depth – although it feels a hundred times longer because of the microscope's magnification – you will find the tumour.

I am looking directly into the centre of the brain, a secret and mysterious area where all the most vital functions that keep us conscious and alive are to be found. Above me, like the great arches of a cathedral roof, are the deep veins of the brain – the Internal Cerebral Veins and beyond them the basal veins of Rosenthal and then in the midline the Great Vein of Galen, dark blue and glittering in the light of the microscope. This is anatomy that inspires awe in neurosurgeons.

These veins carry huge volumes of venous blood away from the brain. Injury to them will result in the patient's death. In front of me is the granular red tumour and beneath it the tectal plate of the brainstem, where damage can produce permanent coma. On either side are the posterior cerebral arteries which supply the parts of the brain responsible for vision. Ahead, beyond the tumour, like a door opening into a distant white-walled corridor once the tumour has been removed, is the third ventricle.

There is a fine, surgical poetry to these names which, combined with the beautiful optics of a modern, counter-balanced microscope, makes this one of the most wonderful of neurosurgical operations – if all goes well, that is. On this occasion as I approached the tumour there were several blood vessels in the way that had to be cut – you need to know which can be sacrificed and which cannot. It was as though I had lost all my knowledge and experience. Every time I divided a blood vessel I shook a little with fright, but as a surgeon you learn at an early stage of your career to accept intense anxiety as a normal part of the day's work and to carry on despite it.

An hour and a half into the operation I reached the tumour. I removed a minute fragment to be sent off to the pathology laboratory and I leaned back in my operating chair.

'We'll now have to wait,' I said to Mike with a sigh. It is not easy to break off in the middle of an operation and I sat slumped in my chair, nervous and tense, longing to get on with the operation, hoping that my pathology colleague would report the tumour to be both benign and operable, hoping that the patient would live, hoping that I would be able to tell his wife after the operation that all would be well.

After forty-five minutes I could not stand the delay any longer, pushed my chair away from the operating table and leapt out of it to go to the nearest phone, still in my sterile gown and gloves. I rang the path lab and demanded to speak to the pathologist. There was a brief delay and he came to the phone.

'The frozen!' I shouted 'What's happening?'

'Ah,' said the pathologist, sounding quite imperturbable. 'So sorry about the delay. I was in another part of the building.'

'What the hell is it?'

'Yes. Well, I'm looking at it now. Ah! Yes, it looks like a straightforward benign pineocytoma ...'

'Wonderful! Thank you!'

Instantly forgiving him, I went back to the operating table where everybody was waiting.

'Let's get on with it!'

I scrubbed up again and climbed back into my operating chair, settled my elbows on the armrests and got back to work on the tumour. Each brain tumour is different. Some are as hard as rock, some as soft as jelly. Some are completely dry, some pour with blood – sometimes to such an extent that the patient can bleed to death during the operation. Some shell out like peas from a pod, others are hopelessly stuck to the brain and its blood vessels. You can never know for certain from a brain scan exactly how a tumour will behave until you start to remove it. This man's tumour was, as surgeons say, cooperative and with a good surgical plane – in other words, it was not stuck to the brain. I slowly cored it out, collapsing the tumour in on itself away from the surrounding brain. After three hours it looked as though I had got most of it out.

Since pineal tumours are so rare one of my colleagues came into my theatre from his own operating theatre, to see how the operation was going. He was probably a little jealous.

He peered over my shoulder.

'Looks OK.'

'So far,' I said.

'Things only go wrong when you're not expecting them,' he replied as he turned to go back to his own theatre.

The operation continued until I had removed all of the tumour without injuring any of the surrounding vital architecture of the brain. I left Mike to close the wound and walked to the wards. I had only a few in-patients, one of them the young mother I had left paralysed a week earlier. I found her on her own in a side-room. When you approach a patient you have damaged it feels as though there is a force-field pushing against you, resisting your attempts to open the door behind which the patient is lying, the handle of which feels as though it were made of lead, pushing you away from the patient's bed, resisting your attempts to raise a hesitant smile. It is hard to know what role to play. The surgeon is now a villain and perpetrator, or at best incompetent, no longer heroic and all-powerful. It is much easier to hurry past the patient without saying anything.

I went into the room and sat down in the chair beside her.

'How are you?' I asked lamely.

She looked at me and grimaced, pointing wordlessly with her good left arm to her paralysed right arm and then lifting it up to let it fall lifeless onto the bed.

'I've seen this happen after surgery before, and the patients got better, although it took months. I really do believe you will get largely better.'

'I trusted you before the operation,' she said. 'Why should I trust you now?'

I had no immediate reply to this and stared uncomfortably at my feet.

'But I believe you,' she said after a while, although perhaps only out of pity.

I went back to the theatres. The pineal patient had been transferred from the table to a bed and was already awake. He lay with his head on a pillow, looking bleary-eyed, while one of the nurses washed the blood and bone dust left from the operating out of his hair. The anaesthetists and theatre staff were laughing and chatting as they busied themselves around him, rearranging the many tubes and cables attached to him, in preparation for wheeling him round to the ITU. If he had not woken up so well they would have been working in silence. The nurses were tidying the instruments on the trolleys and stuffing the discarded drapes and cables and tubes into plastic rubbish bags. One of the porters was already mopping the blood off the floor in preparation for the next case.

'He's fine!' Mike happily shouted to me across the room.

I went to find his wife. She was waiting in the corridor outside the ITU, her face rigid with fear and hope as she watched me approach her.

'It went as well we could hope,' I said, in a formal and matter-of-fact voice, playing the part of a detached and brilliant brain surgeon. But then I could not help but reach out to her, to put my hands on her shoulders, and as she put her hands on mine and we looked into each other's eyes, and I saw her tears and had to struggle for a moment to control my own, I allowed myself a brief moment of celebration.

'I think everything's going to be all right,' I said.

2

ANEURYSM

n. a morbid dilatation of the wall of a blood vessel, usu. an artery.

Neurosurgery involves the surgical treatment of patients with diseases and injuries of the brain and spine. These are rare problems so there are only a small number of neurosurgeons and neurosurgical departments in comparison to other medical specialties. I never saw any neurosurgery as a medical student. We were not allowed into the neurosurgical theatre in the hospital where I trained – it was considered too specialized and arcane for mere students. Once, when walking down the main theatre corridor, I had had a brief view through the small port-hole window of the neurosurgical theatre's door of a naked woman, anaesthetized, her head completely shaven, sitting bolt upright on a special operating table. An elderly and immensely tall neurosurgeon, his face hidden by a surgical facemask and a complicated headlight fixed to his head, was standing behind her. With enormous hands he was painting her bare scalp with dark brown iodine antiseptic. It looked like a scene from a horror film.

Three years later I found myself in that same neurosurgical operating theatre, watching the younger of the two consultant neurosurgeons who worked in the hospital, operating on a woman with a ruptured cerebral aneurysm. I had been qualified as a doctor for one and a half years by then and was already disappointed and disillusioned with the thought of a career in medicine. I was working at the time as a senior house officer, or SHO for short, in my teaching hospital's intensive care unit. One of the anaesthetists who worked on the ITU, seeing that I looked a little bored, had suggested that I come down to the operating theatre to help her prepare a patient for a neurosurgical operation.

It was unlike any other operation I had seen, which had usually seemed to involve long, bloody incisions and the handling of large and slippery body parts. This operation was done with an operating microscope, through a small opening in the side of the woman's head using only fine microscopic instruments with which to manipulate her brain's blood vessels.

Aneurysms are small, balloon-like blow-outs on the cerebral arteries that can – and often do – cause catastrophic haemorrhages in the brain. The aim of the operation is to place a minute spring-loaded metal clip across the neck of the aneurysm – just a few millimetres across – to prevent the aneurysm bursting. There is a very real danger that the surgeon, working at several inches' depth in the centre of the patient's head, in a narrow space beneath the brain, will inadvertently burst the aneurysm while he dissects it free from the surrounding brain and blood vessels and tries to clip it. Aneurysms have thin, fragile walls, yet they have high pressure, arterial blood within them. Sometimes the wall is so thin that you can see the swirling dark red vortices of blood within the aneurysm, made enormous and sinister by the magnification of the operating microscope. If the surgeon ruptures the aneurysm before he can clip it the patient will usually die, or at least suffer a catastrophic stroke – a fate that can easily be worse than death.

The staff in the theatre were silent. There was none of the usual chatter and talk. Neurosurgeons sometimes describe aneurysm surgery as akin to bomb disposal work, though the bravery required is of a different kind as it is the patient's life that is at risk and not the surgeon's. The operation I was watching was more like a blood sport than a calm and dispassionate technical exercise, with the quarry a danger-ous aneurysm. There was the chase – the surgeon cautiously stalking his way beneath the patient's brain towards the an-eurysm, trying not to disturb it, to where it lay deep within the brain. And then there was the climax, as he caught the aneurysm, trapped it, and obliterated it with a glittering, spring-loaded titanium clip, saving the patient's life. More than that, the operation involved the brain, the mysterious substrate of all thought and feeling, of all that was important in human life – a mystery, it seemed to me, as great as the stars at night and the universe around us. The operation was elegant, delicate, dangerous and full of profound meaning. What could be finer, I thought, than to be a neurosurgeon? I had the strange feeling that this was what I had wanted to do all my life, even though it was only now that I had realized it. It was love at first sight.

The operation went well. The aneurysm was successfully clipped without causing a catastrophic stroke or haemorrhage and the atmosphere in the operating theatre was suddenly happy and relaxed. I went home that night and announced to my wife that I was going to be a brain surgeon. She looked a little surprised, given that I had been so undecided about what sort of doctor I should be, but she seemed to think the idea made sense. Neither of us could have known then that my obsession with neurosurgery and the long working hours and the self-importance it produced in me would lead to the end of our marriage twenty-five years later.

*

Thirty years and several hundred aneurysm operations later, re-married and only a few years away from retirement, I cycled in to work on a Monday morning with an aneurysm to clip. A heat wave had just ended and heavy grey rain clouds hung over south London. It had poured with rain during the night. There was little traffic – almost everybody seemed to be away on holiday. The gutters at the entrance to the hospital were flooded so that the passing red buses sent cascades of water over the pavement and the small number of staff walking to work had to jump to one side as the buses swept past.

I rarely clip aneurysms now. All the skills that I slowly and painfully acquired to become an aneurysm surgeon have been rendered obsolete by technological change. Instead of open surgery, a catheter and wire is passed through a needle in the patient's groin into the femoral artery and fed upwards into the aneurysm by a radiology doctor – not a neurosurgeon – and the aneurysm is blocked off from the inside rather than clipped off from the outside. It is, without a doubt, a much less unpleasant experience for patients than being subjected to an operation. Although neurosurgery is no longer what it once was, the neurosurgeon's loss has been the patient's gain. Most of my work is now concerned with tumours of the brain – tumours with names like glioma or meningioma or neurinoma – the suffix '-oma' coming from the ancient Greek word for tumour and the first part of the word being the name of the type of cell from which the tumour is thought to have grown. Occasionally an aneurysm cannot be coiled, so every so often I find myself going to work in the morning in that state of controlled anxiety and excitement that I knew so well in the past.

The morning always starts with a meeting – a practice I began twenty years ago. I had been inspired by the TV police soap *Hill Street Blues*, where every morning the charismatic station police sergeant would deliver pithy homilies and instructions to his officers before they set off onto the city streets in their police cars with their sirens wailing. It was at the time when the government was starting to reduce the long working hours of junior hospital doctors. The doctors were tired and overworked, it was said, and patients' lives were being put at risk. The junior doctors, however, rather than becoming ever more safe and efficient now that they slept longer at night, had instead become increasingly disgruntled and unreliable. It seemed to me that this had happened because they were now working in shifts and had lost the sense of importance and belonging that came with working the long hours of the past. I hoped that by meeting every morning to discuss the latest admissions, to train the juniors with constant teaching as well as to plan the patients' treatment, we might manage to recreate some of the lost regimental spirit.

The meetings are very popular. They are not like the dull and humourless hospital management meetings where there is talk of keeping in the loop about the latest targets or of feeling comfortable about the new Care Pathways. Our neurosurgical morning meeting is a different sort of affair. Every day at eight o'clock sharp, in the dark and windowless X-ray viewing room, we shout and argue and laugh while looking at the brain scans of our poor patients and crack black jokes at their expense. We sit in a semi-circle, a small group of a dozen or so consultants and junior doctors, looking as though we were on the deck of the Starship Enterprise.

Facing us is a battery of computer monitors and a white wall onto which brain scans are projected, many times larger than life-size, in black and white. The scans are of patients admitted as emergencies over the preceding twenty-four hours. Many of the patients will have suffered fatal haemorrhages or severe head injuries, or have newly diagnosed brain tumours. We sit there, alive and well and happy in our work, and with sardonic amusement and Olympian detachment we examine these abstract images of human suffering and disaster, hoping to find interesting cases on which to operate. The junior doctors present the cases, giving us the 'history' as it is called – the stories of sudden catastrophe or of terrible tragedy that are repeated each day, year in, year out, as though human suffering would never end.

I sat down in my usual place at the back, in the corner. The SHOs are in the front row and the surgical trainees, the specialist registrars, sit in the row behind them. I asked which of the junior doctors had been on call for the emergency admissions.

'A locum,' one of the registrars replied, 'and he's buggered off.'

'There were five doctors holding the on-call bleep over twenty-four hours on Friday,' one of my colleagues said. 'Five doctors! Handing over emergency referrals to each other every four point two hours! It's utter chaos ...'

'Is there anything to present?' I asked. One of the juniors got up from his chair and walked to the computer keyboard on the desk at the front of the room.

'A thirty-two-year-old woman,' he said tersely. 'For surgery today. Had some headaches and had a brain scan.' As he talked a brain scan flashed up on the wall.

I looked at the young SHOs and to my embarrassment could not remember any of their names. When I became a consultant twenty-five years ago the department had just two SHOs, now there are eight. In the past I used to get to know them all as individuals and take a personal interest in their careers, but now they come and go as quickly as the patients. I asked one of them to describe the scan on the wall in front of us, apologizing for not knowing who she was.

'Alzheimer's!' one of the less deferential registrars shouted from the darkness at the back of the room.

The SHO told me that she was called Emily. 'This is a CTA of the brain,' she said.

'Yes, we can all see that. But what does it show?'

There was an awkward silence.

After a while I took pity on her. I walked up to the wall and pointed to the scan. I explained how the arteries to the brain were like the branches of a tree, narrowing as they spread outwards. I pointed to a little swelling, a deadly berry, coming off one of the cerebral arteries and looked enquiringly at Emily.

'Is it an aneurysm?' Emily asked.

'A right middle cerebral artery aneurysm,' I replied. I explained how the woman's headaches had in fact been quite mild and the aneurysm was coincidental and had been discovered by chance. It had nothing to do with her headaches.

'Who's doing the exam next?' I asked, turning to look at the row of specialist registrars who all have to take a nationally organized exam in neurosurgery as they reach the end of their training. I try to grill them regularly in preparation for it.

'It's an unruptured aneurysm, seven millimetres in size,' Fiona – the most experienced of the registrars – said. 'So there's a point zero five per cent risk of rupture per year according to the international study published in 1998.'

'And if it ruptures?'

'Fifteen per cent of people die immediately and another thirty per cent die within the next few weeks, usually from a further bleed and then there's a compound interest rate of four per cent per year.'

'Very good, you know the figures. But what should we do?'

'Ask the radiologists if they can coil it.'

'I've done that. They say they can't.'

The interventional radiologists – the specialist X-ray doctors who now usually treat aneurysms – had told me that the aneurysm was the wrong shape and would have to be surgically clipped if it was to be treated.

'You could operate ...'

'But should I?'

'I don't know.'

She was right. I didn't know either. If we did nothing the patient might eventually suffer a haemorrhage which would probably cause a catastrophic stroke or kill her. But then she might die years away from something else without the aneurysm ever having burst. She was perfectly well at the moment, the headaches for which she had had the scan were irrelevant and had got better. The aneurysm had been discovered by chance. If I operated I could cause a stroke and wreck her – the risk of that would probably be about four or five per cent. So the acute risk of operating was roughly similar to the life-time risk of doing nothing. Yet if we did nothing she would have to live with the knowledge that the aneurysm was sitting there in her brain and might kill her any moment.

'So what should we do?' I asked.

'Discuss it with her?'

I had first met the woman a few weeks earlier in my outpatient clinic. She had been referred by the GP who had organized the brain scan but his referral note told me nothing about her other than that she was thirty-two years old and had an unruptured aneurysm. She came on her own, smartly dressed, with a pair of sunglasses pushed back over her long dark hair. She sat down on the chair beside my desk in the dull outpatient room and put her elaborate designer bag down on the floor beside her chair. She looked anxiously at me.

I apologized for keeping her waiting and hesitated before continuing. I did not want to start the interview by immediately asking her about her family circumstances or about herself – it would sound as though I was expecting her to die. I asked her about the headaches.

So she told me about them, and also the fact that they were already better. They certainly sounded harmless in retrospect. If headaches have a serious cause it is usually obvious from the nature of the headaches. The investigation organized by her GP – hoping, perhaps, that a normal brain scan would reassure her – had created an entirely new problem and the woman, although no longer suffering with headaches, was now desperate with anxiety. She had been on the Internet, inevitably, and now believed that she had a time bomb in her head which was about to explode any minute. She had been waiting several weeks to see me.

I showed her the angiogram on the computer on the desk in front of us. I explained that the aneurysm was very small and might very well never burst. It was the large ones which were dangerous and definitely needed treating, I said. I told her that the risks of the operation were probably very much the same as the risk of her having a stroke from the aneurysm bursting.

'Does it have to be an operation?' she asked.

I told her that if she was to be treated it would indeed have to be surgery. The problem was knowing whether to do it or not.

'What are the risks of the operation?' She started to cry as I told her that there was a four to five per cent chance she would die or be left disabled by the operation.

'And if I don't have the operation?' she asked through her tears.

'Well, you might manage to die from old age without the aneurysm having ever burst.'

'They say you're one of the best neurosurgeons in the country,' she said with the naive faith that anxious patients use to try to lessen their fears.

'Well, I'm not. But I'm certainly very experienced. All I can do is promise to do my best. I'm not denying that I'm completely responsible for what happens to you but I'm afraid it's your decision as to whether to have the op or not. If I knew what to do I promise I would tell you.'

'What would you do if it was you?'

I hesitated, but the fact of the matter was that by the age of sixty-one I was well past my best-by date and I knew that I had already lived most of my life. Besides, the difference in our ages meant that I had fewer years of life ahead of me so the life-time risk of the aneurysm rupturing, if it was not operated on, would be much lower for me and the relative risk of the operation correspondingly higher.

'I would not have the aneurysm treated,' I said, 'although I would find it quite hard to forget about it.'

'I want the op,' she said. 'I don't want to live with this thing in my head,' emphatically pointing at her head.

'You don't have to decide now. Go home and talk things over with your family.'

'No, I've decided.'

I said nothing for a while. I was not at all sure she had really listened to what I had told her about the risks of surgery. I doubted if going over it all over again would achieve much so we set off on the long trek along the hospital corridors to find my secretary's office and arrange a date for the operation.

On a Sunday evening three weeks later I trudged in to the hospital, as usual, to see her and the other patients due to have surgery the next day. I went to the hospital reluctantly, irritable and anxious, much of the day having been overhung with the thought of having to see the woman and face her anxiety.

Every Sunday evening I cycle to the hospital full of fore-boding. It is a feeling that seems to be generated merely by the transition from being at home to being at work irrespec-tive of the difficulty of the cases awaiting me. This evening visit is a ritual I have performed for many years and yet, try as I might, I cannot get used to it and escape the dread and pre-occupation of Sunday afternoons – almost a feeling of doom – as I cycle along the quiet backstreets. Once I have seen the patients, however, and spoken to them, and discussed with them what will happen to them next day, the fear leaves me and I return home happily enough, ready for the next day's operating.

I found her in one of the crowded bays on the wom-en's ward. I had hoped her husband might be with her so that I could talk to them together but she told me that he had already left as their children were at home. We talked about the operation for a few minutes. The decision was now made, so I did not feel the need to stress the risks as I had done in the outpatient clinic, although I still had to refer to them when I got her to sign the complicated consent form.

'I hope you get some sleep,' I said. 'I promise you I will, which is more important in the circumstances.' She smiled at the joke – a joke I make with all my patients when I see them the night before surgery. She probably knew already that the last thing you get in hospital is peace, rest or quiet, especially if you are to undergo brain surgery next morning.

I saw the other two patients who were also on the list for surgery and went over the details of their operations with them. They signed the consent forms and as they did so both of them had told me how they trusted me. Anxiety might be contagious, but confidence is also contagious, and as I walked to the hospital car park I felt buoyed up by my patients' trust. I felt like the captain of a ship – everything was in order, everything was ship-shape and the decks were cleared for action, ready for the operating list tomorrow. Playing with these happy nautical metaphors as I left the hospital, I went home.

After the morning meeting I went to the anaesthetic room where the patient was lying on a trolley, waiting to be anaesthetized.

'Good morning,' I said, attempting to sound cheerful. 'Did you sleep well?'

'Yes,' she replied calmly. 'I had a good night's sleep.'

'Everything's going to be fine,' I said.

I could only wonder once again whether she really appreciated the risks to which she was about to be exposed. Perhaps she was very brave, perhaps naive, perhaps she had not really taken in what I had told her.

In the changing room I stripped off and climbed into theatre pyjamas. One of my consultant colleagues was getting changed as well and I asked him what was on his list for the day.

'Oh, just a few backs,' he said 'You've got the aneurysm?'

'The trouble with unruptured aneurysms,' I said, 'is that if they wake up wrecked you have only yourself to blame. They're in perfect nick before the op. At least with the ruptured ones they're often already damaged by the first bleed.'

'True. But the unruptured ones are usually much easier to clip.'

I went in to the theatre where Jeff, my registrar, was positioning the woman on the operating table. My department is unusual in having American surgeons from the neurosurgical training programme in Seattle who train with us for a year at a time. Jeff was one of these and, as with most of the American trainees, he was outstanding. He was clamping her head to the table – three pins attached to a hinged frame are driven through the scalp into the skull to hold the patient's head immobile.

I had promised her a minimal head shave and Jeff started to shave the hair from her forehead. There is no evidence that the complete head shaves we did in the past, which made the patients look like convicts, had any effect on infection rates, which had been the ostensible reason for doing them. I suspect the real – albeit unconscious – reason was that dehumanizing the patients made it easier for the surgeons to operate.

With the minimal head shave completed we go to the scrub-up sink and wash our hands and then, gloved and masked and gowned, return to the table and start the operation. The first ten minutes or so are spent painting the patient's head with antiseptic, covering her with sterile towels so that I can only see the area to be operated upon, and setting up the surgical equipment and instruments with the scrub nurse.

'Knife,' I say to Irwin, the scrub nurse. 'I'm starting,' I shout to the anaesthetist at the other end of the table, and off we go.

After thirty minutes of working with drills and cutters powered by compressed air the woman's skull is open and the uneven ridges of bone on the inside of her skull have been smoothed down with a cutting burr.

'Lights away, microscope in and the operating chair!' I shout, as much from excitement as from the need to make myself heard above the rattle and hum and hissing of all the equipment and machinery in the theatre.

Modern binocular operating microscopes are wonderful things and I am deeply in love with the one I use, just as any good craftsman is with his tools. It cost over one hundred thousand pounds and although it weighs a quarter of a ton it is perfectly counter-balanced. Once in place, it leans over the patient's head like an inquisitive, thoughtful crane. The binocular head, through which I look down into the patient's brain, floats as light as a feather on its counter-balanced arm in front of me, and the merest flick of my finger on the controls will move it. Not only does it magnify, but it illuminates as well, with a brilliant xenon light source, as bright as sunlight.

Two of the theatre nurses, bent over with the effort, slowly push the heavy microscope up to the table and I climb into the operating chair behind it – a specially adjustable chair with armrests. This moment still fills me with awe. I have not yet lost the naive enthusiasm with which I watched that first aneurysm operation thirty years ago. I feel like a medieval knight mounting his horse and setting off in pursuit of a mythical beast. And the view down the microscope into the patient's brain is indeed a little magical – clearer, sharper and more brilliant than the world outside, the world of dull hospital corridors and committees and management and paperwork and protocols. There is an extraordinary sense of depth and clarity produced by the microscope's hugely expensive optics, made all the more intense and mysterious by my anxiety. It is a very private view, and although the surgical team is around me, watching me operate on a video monitor connected to the microscope, and although my assistant is beside me, looking down a side-arm, and despite all the posters in the hospital corridors about something called clinical governance proclaiming the importance of team-working and communication, for me this is still single combat.

'Well, Jeff, let's get on with it. And let's have a brain retractor,' I add to Irwin.

I choose one of the retractors – a thin strip of flexible steel with a rounded end like an ice-cream stick – and place it under the frontal lobe of the woman's brain. I start to pull the brain upwards away from the floor of the skull – *elevation* is the proper surgical word – cautious millimetre by cautious millimetre, creating a narrow space beneath the brain along which I now crawl towards the aneurysm. After so many years of operating with the microscope it has become an extension of my own body. When I use it it feels as though I am actually climbing down the microscope into the patient's head, and the tips of my microscopic instruments feel like the tips of my own fingers.

I point out the carotid artery to Jeff and ask Irwin for the microscopic scissors. I carefully cut the gossamer veil of the arachnoid around the great artery that keeps half the brain alive. The arachnoid, a fine layer of the meninges, is named after the Greek word for a spider, as it looks as though it was made from the strands of the finest spider's web.

'What a fantastic view!' says Jeff. And it is, because we are operating on an aneurysm before a catastrophic rupture and the cerebral anatomy is clean and perfect.

'Let's have another retractor,' I say.

Armed now with two retractors I start to prise apart the frontal and temporal lobes, held together by the arachnoid. Cerebro-spinal fluid, known to doctors as CSF, as clear as liquid crystal, circulating through the strands of the arachnoid, flashes and glistens like silver in the microscope's light. Through this I can see the smooth yellow surface of the brain itself, etched with minute red blood vessels – arterioles – which form beautiful branches like a river's tributaries seen from space. Glistening, dark purple veins run between the two lobes leading down towards the middle cerebral artery and, ultimately, to where I will find the aneurysm.

'Awesome!' Jeff says again.

'CSF used to be called "gin-clear" when there was no

blood or infection in it,' I say to Jeff. 'But probably we're now supposed to use alcohol-free terminology.'

I soon find the right middle cerebral artery. In reality only a few millimetres in diameter, it is made huge and menacing by the microscope – a great pink-red trunk of an artery which ominously pulses in time with the heart-beat. I need to follow it deep into the cleft – known as the Sylvian fissure – between the two lobes of the brain – to find the aneurysm in its lair, where it grows off the arterial trunk. With ruptured aneurysms this dissection of the middle cerebral artery can be a slow and tortuous business, since recent haemorrhage often causes the sides of the two lobes to stick together. Dissecting them is difficult and messy, and there is always the fear that the aneurysm will rupture again while I am doing this.

I separate the two lobes of the brain by gently stretching them apart, cutting the minute strands of arachnoid that bind them together with a pair of microscope scissors in one hand while I keep the view clear of spinal fluid and blood with a small sucker. The brain is a mass of blood vessels and I must try to avoid tearing the many veins and minute arteries both to prevent bleeding from obscuring the view and also for fear of damaging the blood supply to the brain. Sometimes, if the dissection is particularly difficult and intense, or dangerous, I will pause for a while, rest my hands on the arm-rests, and look at the brain I am operating on. Are the thoughts that I am thinking as I look at this solid lump of fatty protein covered in blood vessels really made out of the same stuff? And the answer always comes back – they are – and the thought itself is too crazy, too incomprehensible, and I get on with the operation.

Today, the dissection is easy. It is as though the brain unzips itself, and only the most minimal manipulation is required on my part for the frontal and temporal lobes to part rapidly, so that within a matter of minutes we are looking at the aneurysm, entirely free from the surrounding brain and the dark purple veins, glittering in the brilliant light of the microscope.

'Well. It's just asking to be clipped, isn't it?' I say to Jeff, suddenly happy and relaxed. The greatest risk is now past. With this kind of surgery, if the aneurysm ruptures before you reach it, it can be very difficult to control the bleeding. The brain suddenly swells and arterial blood shoots upwards, turning the operative site into a rapidly rising whirlpool of angry, swirling red blood, through which you struggle desperately to get down to the aneurysm. Seeing this hugely magnified down the microscope you feel as though you are drowning in blood. One quarter of the blood from the heart goes to the brain – a patient will lose several litres of blood within a matter of minutes if you cannot control the bleeding quickly. Few patients survive the disaster of premature rupture.

'Let's have a look at the clips,' I say.

Irwin hands me the metal tray containing the gleaming titanium aneurysm clips. They come in all shapes and sizes, corresponding to the many shapes and sizes of aneurysms. I look at the aneurysm down the microscope and at the clips and then back at the aneurysm.

'Six millimetre, short right-angled' I tell him.

He picks out the clip and loads it onto the applicator. The applicator consists of a simple instrument with a handle formed by two curved leaf springs, joined at either end. Once the clip is loaded at the instrument's tip, all you have to do is press the springs of the handle together to open the blades of the clip, position the opened blades carefully across the neck of the aneurysm and then allow the springs to separate gently apart within your hand so that the clip blades close across the aneurysm, sealing it off from the artery from which it has grown, so that blood can no longer get into it. By finally letting the springs of the handle separate even more fully the clip is released from the applicator which you can then withdraw, leaving the clip clamped across the aneurysm for the rest of the patient's life.

That, at least, is what is supposed to happen and had always happened with the hundreds of similar operations I had carried out in the past.

Since this looks a straightforward aneurysm to clip I let Jeff take over, and I clamber out of the operating chair so that he can replace me. My assistants are all as susceptible to the siren call of aneurysms as I am. They long to operate on them, but the fact that most aneurysms are now coiled rather than clipped means that it is no longer possible to train them properly and I can only give them the simplest and easiest parts of the occasional operation to do, under very close supervision.

Once Jeff is settled in, the nurse hands him the loaded clip applicator, and he cautiously moves it towards the aneurysm. Nothing much seems to happen, and down the assistant's arm of the microscope I nervously watch the clip wobble uncertainly around the aneurysm. It is a hundred times more difficult and nerve-wracking to train a junior surgeon than it is to operate oneself.

After a while – probably only a few seconds though it feels much longer – I can stand it no longer.

'You're fumbling. I'm sorry but I'll have to take over.'

Jeff says nothing and climbs out of the chair – it would be a rash surgical trainee who ever complained to his boss, especially at a moment like this – and we change places again.

I take the applicator and place it against the aneurysm, pressing the springs of the handle together. Nothing happens.

'Bloody hell, the clip won't open!'

'That was the problem I was having,' Jeff says, sounding a little aggrieved.

'Bloody hell! Well, give me another applicator.'

This time I easily open the clip and slip the blades over the aneurysm. I open my hand and the blades close, neatly clipping the aneurysm. The aneurysm, defeated, shrivels since it is now no longer filling with high pressure arterial blood. I sigh deeply – I always do when the aneurysm is finally dealt with. But to my horror I find that this second applicator has an even more deadly fault than the first: having closed the clip over the aneurysm the applicator refuses to release the clip. I cannot move my hand for fear of tearing the minute, fragile aneurysm off the middle cerebral artery and causing a catastrophic haemorrhage. I sit there motionless, with my hand frozen in space. If an aneurysm is torn off its parent artery you can usually only stop the bleeding by sacrificing the artery, which will result in a major stroke.

I swear violently while trying to keep my hand steady.

'What the fuck do I do now?' I shout to no one in particular. After a few seconds – it feels like minutes – I realize that I have no choice other than to remove the clip, despite the risk that this might cause the aneurysm to burst. I re-close the applicator handle and to my relief the blades of the clip open easily. The aneurysm suddenly swells and springs back into life, filling instantly with arterial blood. I feel it is laughing at me and about to burst but it doesn't. I throw myself back in my chair, cursing even more violently, and then hurl the offending instrument across the room.

'That's never happened before!' I shout but then, quickly calming down, laugh to Irwin, 'And that's only the third time in my career I've thrown an instrument onto the floor.'

I have to wait a few minutes while yet another applicator was found. The faulty ones, for some strange reason, turned out to have stiff hinges. Only later did I remember that the surgeon I had watched thirty years ago, and whose trainee I became, had told me that he had once encountered the same problem, although his patient had been less fortunate than mine. He was the only surgeon I knew who always checked the applicator before using it.

Doctors like to talk of the 'art and science' of medicine. I have always found this rather pretentious, and prefer to see what I do as a practical craft. Clipping aneurysms is a skill, and one that takes years to learn. Even when the aneurysm is exposed and ready to take a clip, after the thrill of the chase, there is still the critical question of how I place the clip across the aneurysm, and the all-important question of whether I have clipped the aneurysm's neck completely without damaging the vital artery from which the aneurysm has grown.

This aneurysm looks relatively easy but my nerves are too frayed to let my assistant take over again and so, with yet another applicator, I clip the aneurysm. The shape of this aneurysm, however, is such that the clip does not pass completely over the neck – I can just see a little part of the aneurysm neck sticking out beyond the tips of the clip.

'Not quite across,' Jeff says helpfully.

'I know!' I snap.

This is a difficult part of the operation. I can partly open the clip and re-position it to get a more perfect position but I might tear the aneurysm in the process and be left looking at a fountain of arterial blood rushing up the microscope towards me. On the other hand if the aneurysm neck is not completely occluded there is some danger – though it is difficult to say how great – that the patient will eventually suffer a further haemorrhage in the future.

A famous English surgeon once remarked that a surgeon has to have nerves of steel, the heart of a lion and the hands of a woman. I have none of these and instead, at this point of an aneurysm operation, I have to struggle against an overwhelming wish to get the operation over and done with, and to leave the clip in place, even if it is not quite perfectly placed.

'The best is the enemy of the good,' I will growl at my assistants, for whom the operation is a wonderful spectator sport. They take a certain pleasure in pointing out that I have not clipped the aneurysm as well as I might have done, since they will not have to cope with the consequences of the aneurysm tearing. And if that happens, it is always exciting to watch their boss struggling with torrential haemorrhage – I certainly enjoyed it when I was a trainee. Besides they will not have to experience the hell of seeing the wrecked patient afterwards on the ward round and feel responsible for the catastrophe.

'Oh, very well,' I will say, shamed by my assistant, but also thinking of the hundreds of aneurysms I have clipped in the past and how, like most surgeons, I have become bolder with experience. Inexperienced surgeons are too cautious – only with endless practice do you learn that you can often get away with things that at first seemed far too frightening and difficult.

I cautiously open the clip a little and gently push it further along the aneurysm.

'There's still a little bit out,' says Jeff.

Sometimes at these moments my past disasters with aneurysm surgery parade before me like ghosts. Faces, names, wretched relatives I forgot years ago suddenly reappear. As I struggle against my urge to finish the operation and escape the fear of causing a catastrophic haemorrhage, I decide at some unconscious place within myself, where all the ghosts have assembled to watch me, whether to re-position the clip yet again or not. Compassion and horror are balanced against cold, technical precision.

I re-position the clip a third time. It finally looks well placed.

'That will do,' I say.

'Awesome!' says Jeff happily, but sad not to have put the clip on himself.

I left Jeff to close, retired to the surgical sitting room next to the theatre and lay down on the large red leather sofa which I had bought for the room some years ago and thought, once again, of how so much of what happens to us in life is determined by random chance. After brain surgery all patients are woken up quickly by the anaesthetist so that we can see if they have suffered any harm or not. With difficult operations all neurosurgeons will wait anxiously for the anaesthetic to be reversed, even if – as with this operation – one is fairly certain that no harm has been done. She awoke perfectly, and once I had seen her I left the hospital to go home.

As I cycled away from the hospital under dull, grey clouds, perhaps I felt only a little of the joy that I used to feel in the past after successful aneurysm operations. At the end of a successful day's operating, when I was younger, I felt an intense exhilaration. As I walked round the wards after an operating list with my assistants beside me and received my patients' heart-felt gratitude and that of their families, I felt like a conquering general after a great battle. There have been too many disasters and unexpected tragedies over the years, and I have made too many mistakes for me to experience such feelings now, but I still felt pleased with the way the operation had gone. I had avoided disaster and the patient was well. It was a deep and profound feeling which I suspect few people other than surgeons ever get to experience. Psychological research has shown that the most reliable route to personal happiness is to make others happy. I have made many patients very happy with successful operations but there have been many terrible failures and most neurosurgeons' lives are punctuated by periods of deep despair.

I went back into the hospital that evening to see the woman. She was sitting up in bed, with the large black eye and swollen forehead that many patients have for a few days after an operation like hers. She told me that she felt sick and had a headache. Her husband was sitting beside her and looked angrily at me as I quickly dismissed her bruises and post-operative pain. Perhaps I should have expressed more sympathy but after the near disaster of the operation I found it difficult to take her minor post-operative problems seriously. I told her that the operation had been a complete success and that she would soon feel better. I had not had the opportunity to talk to her husband before the operation – something I usually take great care to do with relatives – and he had probably appreciated the risks of the operation even less than his wife had done.

We have achieved most as surgeons when our patients recover completely and forget us completely. All patients are immensely grateful at first after a successful operation but if the gratitude persists it usually means that they have not been cured of the underlying problem and that they fear that they may need us in the future. They feel that they must placate us, as though we were angry gods or at least the agents of an unpredictable fate. They bring presents and send us cards. They call us heroes, and sometimes gods. We have been most successful, however, when our patients return to their homes and get on with their lives and never need to see us again. They are grateful, no doubt, but happy to put us and the horror of their illness behind them. Perhaps they never quite realized just how dangerous the operation had been and how lucky they were to have recovered so well. Whereas the surgeon, for a while, has known heaven, having come very close to hell.

3

HAEMANGIOBLASTOMA

n. a tumour of the brain or spinal cord arising from the blood vessels.

I arrived at work feeling cheerful. There was a solid cerebellar haemangioblastoma on the list. These are rare tumours which are formed of a mass of blood vessels. They are benign – meaning that they can be cured by surgery – but they will prove fatal if untreated. There is a small risk of disaster with surgery, since the mass of blood vessels can cause catastrophic haemorrhage if you do not handle the tumour correctly, but there is a much greater chance of success. This is the kind of operation that neurosurgeons love – a technical challenge with a profoundly grateful patient at the end of it if all goes well.

I had seen the patient in my outpatient clinic a few days earlier. He had been suffering from severe headaches for the last few months. He was a forty-year-old accountant, with a head of curly brown hair and a slightly red face that made him look continually embarrassed. As we spoke, I felt embarrassed in return and became self-conscious and awkward as I tried to explain the gravity of his illness to him. Only later did I realize that he had a red face because he was polycythaemic – he had more red blood cells in his blood than normal, since his particular tumour can stimulate the bone marrow to over-produce red blood cells.

'Do you want to see your brain scan?' I asked him, as I ask all my patients.

'Yes ...' he replied, a little uncertainly. The scan made the tumour look as though it was full of black snakes – 'flow voids' – produced by the blood rushing through the potentially disastrous blood vessels. I viewed these on the scan with enthusiasm, as they meant that a challenging operation was in prospect. My patient looked cautiously at the computer screen in front of us as I explained the scan to him and we discussed his symptoms.

'I've never been seriously ill before,' he said unhappily. 'And now this.'

'I'm almost certain it's benign,' I told him. Many brain tumours are malignant and incurable and I often have to overcome my instinct, when talking to patients with brain tumours, to try to comfort and reassure them – I have sometimes failed to do this and have bitterly regretted being too optimistic before an operation. I told him that if I thought it was benign it almost certainly was. I then delivered my standard speech about the risks of the operation and how they had to be justified by the risks of doing nothing. I said that he would die within a matter of months if he did not have the tumour removed.

'Informed consent' sounds so easy in principle – the surgeon explains the balance of risks and benefits, and the calm and rational patient decides what he or she wants – just like going to the supermarket and choosing from the vast array of toothbrushes on offer. The reality is very different. Patients are both terrified and ignorant. How are they to know whether the surgeon is competent or not? They will try to overcome their fear by investing the surgeon with superhuman abilities.

I told him that there was a one or two per cent risk of

his dying or having a stroke if the operation went badly. In truth, I did not know the exact figure as I have only operated on a few tumours like his – ones as large as his are very rare – but I dislike terrorizing patients when I know that they have to have an operation. What was certain was that the risk of the operation was many times smaller than the risk of not operating. All that really matters is that I am as sure as I can be that the decision to operate is correct and that no other surgeon can do the operation any better than I can. This is not as much of a problem for me now that I have been operating on brain tumours for many years, but it can be a moral dilemma for a younger surgeon. If they do not take on difficult cases, how will they ever get any better? But what if they have a colleague who is more experienced?

If patients were thinking rationally they would ask their surgeon how many operations he or she has performed of the sort for which their consent is being sought, but in my experience this scarcely ever happens. It is frightening to think that your surgeon might not be up to scratch and it is much easier just to trust him. As patients we are deeply reluctant to offend a surgeon who is about to operate on us. When I underwent surgery myself, I found that I was in awe of the colleagues who had to treat me though I knew that they, in turn, were frightened of me as all the usual defences of professional detachment collapse when treating a colleague. It is not surprising that all surgeons hate operating on surgeons.

My patient listened in silence as I told him that if I operated upon one hundred people like him, one or two of them would die or be left hopelessly disabled.

He nodded and said what almost everybody says to me in reply to this: 'Well, all operations have risks.'

Would he have chosen not to have the operation if I had said that the risk was five per cent, or fifteen per cent, or fifty per cent? Would he have chosen to find another surgeon who quoted lower risks? Would he have chosen differently if I had not made any jokes, or had not smiled?

I asked him if he had any questions but he shook his head. Taking the pen I offered him he signed the long and complicated form, printed on yellow paper and several pages in length, with a special section on the legal disposal of body parts. He did not read it – I have yet to find anybody who does. I told him that he would be admitted for surgery the following Monday.

'Sent for the patient?' I asked as I entered the operating theatre on Monday morning.

'No,' said U-Nok the ODA (the member of the theatre team who assists the anaesthetist). 'No blood.'

'But the patient has been in the building for two days already,' I said.

U-Nok, a delightful Korean woman, smiled apologetically but said nothing in reply.

'The bloods had to be sent off again at six this morning,' said the anaesthetist as she entered the room. 'They had to be done again because yesterday's bloods were on the old EPR system which has stopped working for some reason because of the new hospital computer system which went live today. The patient now has a different number apparently and we can't find the results from all the blood tests sent yesterday.'

'When can I start?' I asked, unhappy at being kept waiting when I had a dangerous and difficult case to do. Starting on time, with everything just right, and the surgical drapes placed in exactly the right way, the instruments tidily laid out, is an important way of calming surgical stage fright.

'A couple of hours at least.'

I said that there was a poster downstairs saying that iCLIP, the new computer system should only keep patients waiting a few extra minutes.

The anaesthetist laughed in reply. I left the room. Years ago, I would have stormed off in a rage, demanding that something be done, but my anger has come to be replaced by fatalistic despair as I have been forced to recognize my complete impotence as just another doctor faced by yet another new computer program in a huge, modern hospital.

I found the junior doctors in the theatre corridor standing around the reception desk, where a young man was sitting in front of the receptionists' computer with an embarrassed smile. He wore a white PVC tabard on which was stamped in friendly blue letters, on both back and front, 'iCLIP Floorwalker'.

I looked questioningly at Fiona, my senior registrar.

'We've asked him to find the blood results for the brain tumour case but he's not succeeding,' she said.

'I suppose I should go and apologize to the poor patient,' I said with a sigh. I dislike talking to patients on the morning of their operation. I prefer not to be reminded of their humanity and their fear, and I do not want them to suspect that I, too, am anxious.

'I've already told him,' Fiona replied to my relief.

I left the junior doctors and returned to my office, where my secretary Gail had now been joined by Julia the bed manager, one of our senior nurses, who is responsible for the thankless task of trying to find beds for our patients. There are never enough beds, and she spends her working day on the telephone, frantically trying to cajole other bed managers elsewhere to swap one patient for another or to take patients back from the neurosurgical wards so that we can admit a new one.

'Look!' said Gail. She pointed to the welcome screen for iCLIP that she had opened. I saw bizarre names such as Mortuary Discharge, Reverse Decease or Birth Amendments

– each with its own colourful little icon – flash past as she scrolled through the long list.

'I have got to select from this insane list every time I do anything at all!' said Gail.

I left her to struggle with the strange icons and sat in my office doing paperwork until I was telephoned to be told that the patient had finally arrived in the anaesthetic room.

I went upstairs, changed, and joined Fiona in the operating theatre. The patient, now anaesthetized and unconscious, was wheeled into theatre with a little entourage of two anaesthetists, two porters and U-Nok the ODA, pulling drip stands and monitoring equipment along with a tangle of tubes and cables trailing behind the trolley. His face was now hidden by broad swathes of sticking plaster, protecting his eyes and keeping the anaesthetic gas tubing and facial muscle monitoring wires in position. This metamorphosis from person to object is matched by a similar change in my state of mind. The dread has gone, and has been replaced by fierce and happy concentration.

As the tumour was at the base of the man's brain, and as there was the risk of heavy blood loss, I had decided to carry out the operation in what is called, simply enough, the sitting position. The unconscious patient's head is attached to the pin headrest which in turn is connected to a shiny metal scaffold, attached to the operating table. The table is then split and the top half hinged upwards, so that the patient is sitting bolt upright. This helps reduce blood loss during surgery and also improves access to the tumour, but involves a small risk of anaesthetic disaster as the venous blood pressure in the patient's head in the sitting position is below atmospheric room pressure. If the surgeon tears a major vein air can be sucked into the heart, with potentially terrible consequences. As with all operating, it is a question of balancing risks, sophisticated technology, experience and skill, and of luck. With the anaesthetists, the theatre porters and U-Nok, Fiona and I positioned the patient. It took half an hour to make sure his unconscious form was upright with his head bent forward, that there were no 'pressure points' on his arms or legs where pressure sores might develop, and that all the cables and wires and tubes connected to his body were free and not under tension.

'Well, let's get on with it,' I said.

The operation went perfectly with scarcely any blood loss at all. This type of tumour is the only time in brain tumour surgery that you have to remove the tumour 'en bloc' – in a single piece – since if you enter the tumour you will be instantly faced by torrential bleeding. With all other tumours in brain surgery you gradually 'debulk' it, sucking or cutting out the inside of it, collapsing it in on itself, away from the brain, and thus minimizing damage to the brain. With solid haemangioblastomas, however, you 'develop the plane' between the tumour and the brain, creating a narrow crevice a few millimetres wide by gently holding the brain away from the surface of the tumour. You coagulate and divide the many blood vessels that cross from the brain to the tumour's surface, trying not to damage the brain in the process. All this is done with a microscope under relatively high magnification – although the blood vessels are tiny, they can bleed prodigiously. One quarter of the blood pumped every minute by the heart, after all, goes to the brain. Thought is an energy-intensive process.

If all goes well the tumour is eventually freed from the brain and the surgeon will lift the tumour out of the patient's head.

'All out!' I shout triumphantly to the anaesthetist at the other end of the table, and wave the scruffy and bloody little tumour, no bigger than the end of my thumb, in the air at the end of a pair of dissecting forceps. It hardly looked worth all the effort and anxiety.

With the day's operating finished I went to see the patient on the Recovery Ward. He looked remarkably well and wide awake. His wife was beside him and they expressed their heartfelt gratitude.

'Well, we were lucky,' I said to them, though they probably thought this was false modesty on my part, which I suppose to an extent it was.

As I left, dutifully splashing alcoholic hand gel on my hands on my way out, James the registrar on-call for emergencies came looking for me.

'I think you're the consultant on call today,' he said.

'Am I? Well, what have you got?'

'Forty-six-year-old man with a right temporal clot with intraventricular extension in one of the local hospitals – looks like an underlying AVM. GCS five. He was talking when he was admitted.'

An AVM is an arterio-venous malformation, a congenital abnormality which consists of a mass of blood vessels that can, and often do, cause catastrophic haemorrhages. The GCS is the Glasgow Coma Scale and a way of assessing a patient's conscious level. A score of five meant that the man was in coma, and close to death.

I asked him if he had seen the scan and if the patient was already on a ventilator.

'Yes,' James replied, so I asked him what he wanted to do. He was one of the more senior trainees and I knew that he could deal with this case himself.

'Get him up here quickly,' he said. 'There's a bit of hydrocephalus so I'd stick a wide bore drain in and then take out the clot, leaving the AVM alone. It's deep.'

'Carry on,' I said. 'He's potentially salvageable so make

sure they send him up the motorway pronto. You might point out to the local doctors that there's no point sending him if they don't do it quickly. Apparently they need to use the magic phrase "Time Critical Transfer" with the ambulance service and then they won't mess about.'

'It's already done,' James replied happily.

'Splendid!' I said. 'Just get on with it.' And I headed off downstairs to my office.

I cycled home, stopping off at the supermarket to get some shopping. Katharine, the younger of my two daughters, was staying with me for a few days and was to cook supper. I had agreed to do the shopping. I joined a long queue of people at the check-out.

'And what did *you* do today?' I felt like asking them, annoyed that an important neurosurgeon like myself should be kept waiting after such a triumphant day's work. But I then thought of how the value of my work as a doctor is measured solely in the value of other people's lives, and that included the people in front of me in the check-out queue. So I told myself off and resigned myself to waiting. Besides, I had to admit to myself that soon I will be old and retired and then I will no longer count for much in the world. I might as well start getting used to it.

While I was standing in the queue my mobile phone went off. I experienced an immediate flash of alarm, instantly frightened that this would be my registrar calling to say that there was a problem with the brain tumour case but instead I heard an unfamiliar voice as I scattered my shopping over the counter while struggling to answer the phone.

'Are you the consultant neurosurgeon on call?'

Emergency calls are usually all sent to the on-call registrar so I answered warily.

'Yes?' I said.

43

'I am one of the A&E SHOs,' said the voice self-importantly. 'My consultant has told me to ring you about a patient here. Your on-call registrar is not answering his bleep.'

I was immediately annoyed. If the case was so urgent why didn't the A&E consultant ring me himself? There used to be a certain etiquette about ringing a colleague.

'I find that hard to believe,' I said, as I tried to gather up the hot cross buns and clementines I had dropped. A&E were probably just trying to shift patients quickly to meet their target for waiting times. 'I was just speaking to him ten minutes ago ...'

The A&E SHO didn't seem to be listening.

'It's a sixty-seven-year-old man with an acute on chronic subdural ...' he began.

I interrupted him and told him to ring Fiona, who was not on call but I knew was still in the building and then switched the phone off, giving an apologetic smile to the puzzled check-out girl.

I left the supermarket feeling anxious. Perhaps the patient was desperately ill, perhaps James *had* failed to answer his bleep so I rang Fiona on her mobile. I explained the problem and said that I was worried that maybe just for once it really was an urgent referral and not just an attempt to get a patient out of A&E.

I went home. She rang me half an hour later.

'You wait until you hear this one,' she said, laughing. 'James had answered the call and was already on his way to A&E. The patient was perfectly well, he was eighty-one not sixty-seven and they'd completely misinterpreted the brain scan, which was normal.'

'Bloody targets.'

By the time that I had got home it had started to rain. I changed into my running clothes and reluctantly headed for the small suburban park behind my home. Exercise is supposed to postpone Alzheimer's. After a few laps round the park my mobile phone went off.

'Bloody hell!' I said, dropping the wet and slippery phone as I tried to pull it out of my tracksuit and answer the call.

'James here. I can't stop the oozing,' a voice said from the muddy ground.

'What's the problem?' I asked, once I had managed to pick the phone up.

'I've taken the clot out and put a drain in but the cavity is oozing a lot.'

'Not to worry. Line it with Surgicel, pack it and take a break. Go and have a cup of tea. Tea is the best haemostatic agent! I'll look by in thirty minutes or so.'

So I finished my run, had a shower, and made the short journey back to the hospital, but in my car, because of the rain. It was dark by now, with a strong wind, and there had been heavy snowfalls in the north, even though it was already April. I parked my car in the scruffy delivery bay by the hospital basement. Although I am not supposed to park there, it does not seem to matter at night and it means that I can get up to the theatres more quickly than from one of the official car parks which are further away.

I put my head past the doors of the theatre. James was standing at the end of the operating table, holding the patient's head in his hands as he wound a bandage around it. The front of his gown was smeared with blood and there was a large pool of dark red blood at his feet. The operation was clearly finished.

'All well?' I asked.

'Yes. It's fine,' he replied. 'But it took quite a while.'

'Did you go and have a cup of tea to help stop the bleeding?'

'Well, no, not tea,' he said, pointing to a plastic bottle of Coca-Cola on one of the worktops behind him.

'Well, no wonder the haemostasis took so long!' I said with mock disapproval and all the team laughed, happy that the case was over and that they could now go home. I went briefly to check on the tumour patient who was now on the ITU for the night as a matter of routine.

The ITU had had a busy week and there were ten patients in the large and brightly-lit warehouse of a room, all but one of them unconscious, lying on their backs and attached to a forest of machinery with flashing lights and digital read-outs the colour of rubies and emeralds. Each patient has their own nurse, and in the middle of the room there is a large desk with computer monitors and many members of staff talking on the phone or working on the computers or snatching a plastic cup of tea in between carrying out the constant tasks that are needed in intensive care.

The one patient who was not unconscious was my brain tumour case, who was sitting upright in bed, still looking red-faced, but wide awake.

'How are you feeling?' I asked.

'Fine,' he replied with a tired smile.

'Well done!' I replied, as I think patients need to be con-gratulated for their surviving just as much as the surgeons should be congratulated for doing their job well.

'It's a bit of a war zone here, I'm afraid,' I said to him, gesticulating to the depersonalized forms of the other patients and all the technology and busy staff around us. Few – if any – of these patients would survive or emerge unscathed from whatever it was that had damaged their brains.

'I'm afraid you won't get much sleep tonight.'

He nodded in reply, and I went downstairs to the base-ment in a contented frame of mind.

I found my car with a large notice stuck to the windscreen.

'You have been clamped,' the notice said, and there was a long list beneath this accusing me of negligence and disrespect and so on and so forth, and telling me to report to the Security Office to pay a large fine.

'I really can't take this anymore!' I burst out in rage and despair, shouting at the concrete pillars around me but when I furiously marched round my car, to my surprise I found that none of the wheels had been clamped and then, when I came round to the notice again, I noticed that added in ballpoint to the notice were the words 'Next time' with two large exclamation marks.

I drove home torn between impotent rage and gratitude.

4

MELODRAMA

n. a sensational, dramatic piece with crude appeals to the emotions and usu. a happy ending.

I was recently asked to talk to the script-writing team for the TV medical drama *Holby City*. I took the train from Wimbledon to Borehamwood at the opposite end of London and went to the well-appointed country house hotel where they were meeting. There were at least twenty people sitting round a long table. They were thinking of adding a neurosurgical ward, they told me, to the fictional Holby City General Hospital, and wanted me to talk to them about neurosurgery. I talked for almost an hour without stopping, something I don't find very difficult to do, but I probably concentrated too much on the grim and tragic aspects of my work.

'Surely you have some more positive stories to tell, which our viewers would like?' somebody asked and then I suddenly remembered Melanie.

'Well' I said,' Many years ago I did once operate on a young mother who was just about to have a baby and was going blind ...'

There were three patients for surgery on that Wednesday – two women with brain tumours and a young man with a disc prolapse in his lumbar spine. The first patient was Melanie – a twenty-eight-year-old woman in the thirty-seventh week of pregnancy who had started to go blind over the preceding three weeks. She had been referred as an emergency to my neurosurgical department from the ante-natal clinic of her local hospital on Tuesday afternoon. A brain scan had shown a tumour. I was on call for emergencies that day so she was admitted under my care. Her husband had driven her to my hospital from the ante-natal clinic; when I first saw them on the Tuesday afternoon he was guiding Melanie down the hospital corridor towards the ward with one hand on her shoulder and the other hand holding a suitcase. She had her right arm stretched out in front of her for fear of bumping into things and her left hand was pressed onto the unborn child inside her as though she was frightened she might lose it just as she was losing her eyesight. I showed them the way to the ward entrance and said that I would come back later to discuss what should be done.

The brain scan had shown a meningioma – a 'suprasellar' meningioma growing from the meninges, the membrane that encases the brain and spinal cord – at the base of her brain. It was pressing upwards onto the optic nerves where they run back from the eyes to enter the brain. These particular tumours are always benign and usually grow quite slowly, but some of them have oestrogen receptors and, very occasionally, the tumours can expand rapidly during pregnancy when oestrogen levels rise. This was clearly what was happening in Melanie's case. The tumour did not pose a risk to the unborn child, but if it was not removed quickly Melanie would go completely blind. It could happen within a matter of days. An operation to remove a tumour like hers is relatively straightforward but if the visual loss before surgery is severe it is by no means certain it will restore vision and there is some risk it will make it worse. I have once left one person completely blind with a similar operation. Admittedly he was already almost blind before the operation – but then so was Melanie.

When I went to the ward an hour or so later I found Melanie sitting up in her bed, with a nurse beside her completing the admission paperwork. Her husband, looking desperate, was on a chair next to the bed. I sat down on the end of the bed and introduced myself. I asked her how it had all started.

'Three weeks ago. I scraped the side of the car on the garage gates when I was coming home from my ante-natal class,' she said. 'I couldn't understand how I had managed to do it but a few days later I realized that I couldn't see properly out of my left eye.' As she spoke her eyes moved restlessly with the slightly unfocused look that people have when they are going blind. 'It's been getting worse and worse since then,' she added.

'I need to examine your vision a bit,' I said. I asked her if she could see my face.

'Yes,' she replied. 'But it's all blurry.'

I held up my hand in front of her face with the fingers outstretched. I asked her how many fingers she could see.

'I don't really know,' Melanie said with a note of desperation 'I can't see ...'

I had brought an ophthalmoscope, the special torch used for looking into eyes, from my office. I fiddled with the dial on the ophthalmoscope, put my face close up to hers, and focused on the retina of her left eye.

'Look straight ahead,' I said. 'Don't look into the light since it makes your pupil smaller.'

The eyes are said by poets to be the windows to the soul but they are also windows to the brain: examining the retina gives a good idea of the state of the brain as it is directly connected to it. The miniature blood vessels in the eye will be in a very similar condition to the blood vessels in the brain. To my relief I could see that the end of the optic nerve in her eye still looked relatively healthy and not severely damaged, as did the retinal blood vessels. There was some chance surgery would get her better rather than just stop her going completely blind.

'Doesn't look too bad,' I said, after looking into her right eye.

'My baby! What will happen to my baby?' Melanie asked me, clearly more troubled about her child than her eyesight.

I reached out and held her hand and I told her that her baby would be fine. I had already arranged with the obstetricians that they would come and perform a Caesarean section and deliver the baby once I, so to speak, had delivered the tumour. It could all be done under the same anaesthetic, I said. I hoped that surgery would improve her eyesight as well, but had to warn her and her husband that I could not guarantee this. There was also some risk, I told them, that the operation might leave her blind. It was all a question of whether the tumour was badly stuck to the optic nerves or not, which I would not know until I had operated. All that was certain, I said, was that she would go completely blind without surgery. I added that I had seen many patients in poor countries like Ukraine and Sudan who had indeed gone completely blind with tumours such as hers because of delays in treatment. I asked her to sign the consent form. Her husband leant forward and guided her hand with the pen. She scribbled something illegible.

I carried out the operation first thing the next morning with Patrik, the senior registrar who was working with me at the time. The operation had inevitably caused great excitement and there was a small army of obstetricians, paediatricians and nurses with paediatric resuscitation kit in the corridor outside the operating theatre. Doctors and nurses enjoy dramatic cases like this and there was a carnival-like atmosphere to the morning. Besides, the idea of a baby being born in our usually rather grim neurosurgical operating theatres was delightful and the theatre staff were all looking forward to the event as well. The only worry – which was largely mine and Melanie's and her family's – was whether I could save her eyesight or whether I might even leave her completely blind.

She was brought to the theatre from the women's ward on a trolley with her husband walking beside her, her pregnant belly rising up like a small mountain under a hospital sheet. Her husband, fighting back his tears, kissed her goodbye outside the doors to the anesthetic room and was then escorted out of the theatre by one of the nurses. Once Judith had anaesthetized her, Melanie was rolled onto her side and Judith carried out a lumbar puncture, using a large needle up which she then threaded a fine white catheter which we would use to drain all the cerebrospinal fluid out of Melanie's head. This would create more space inside her head – a matter of a few millimetres – in which I could operate.

After a minimal headshave Patrik and I made a long curving incision a centimetre or so behind her hairline following it all the way across her forehead. Pressing firmly with the tips of our fingers on either side of the incision to stop the scalp bleeding we placed plastic clips over the skin edges to close off the skin's blood vessels. We then pulled her scalp off her forehead and folded it down over her face, already covered in the adhesive tape that secured Judith's anaesthetic tube in place. I talked Patrik through the opening stages of the procedure.

'She's young, she's good-looking,' I said. 'We want a good cosmetic result.' I showed him how to make a single burr hole in the skull just out of sight behind the orbit and then use a wire saw called a Gigli saw after its inventor – a sort of glorified cheese wire which makes a much finer cut through bone than the power tools we usually use – to make a very small opening in the skull just above Melanie's right eye. Using the Gigli looks brutal since, as you use your hands to pull the saw backwards and forwards, a fine spray of blood and bone flies upwards and the saw makes an unpleasant grating sound. But, as I said to Patrik, it makes a fine and perfect cut.

Once Patrik had removed the small bone flap – measuring only three centimetres or so – I took over for a while, and used an air-powered drill to smooth off the inside of Melanie's skull. There are a series of ridges, like a microscopic mountain range, two to three millimetres in height, that run across the floor of the skull. By drilling them flat I create a little more space beneath the brain so that I can use less retraction when climbing down under the brain to get at the tumour. I told Patrik to open the meninges with a pair of scissors. The lumbar drain had done its work and the blue-grey dura, the outer layer of the meninges, was shrunken and wrinkled as the brain had collapsed downwards away from the skull as the cerebrospinal fluid had been removed. Patrik tented up the dura with a pair of fine-toothed forceps and started to cut an opening in it with a pair of scissors. Patrik was a short, determined and outspoken Armenian-American.

'They're blunt. They don't cut, they chew,' he said as the scissors jammed on the leathery meninges. 'Give me another pair.' Maria the scrub-up nurse turned back to her trolley and returned with a different pair with which Patrik now exposed the tip of the right frontal lobe of Melanie's brain by cutting through the dura and folding it forwards.

The right frontal lobe of the human brain does not have any specific role in human life that is clearly understood. Indeed, people can suffer a degree of damage to it without seeming to be any the worse for it, but extensive damage will result in a whole range of behavioural problems that are grouped under the phrase 'personality change'. There was little risk of this happening to Melanie but if we damaged the surface of her brain as we lifted the right frontal lobe up by a few millimetres to reach the tumour it was quite likely that we would leave her with life-long epilepsy. It was good to see that Melanie's brain, as a result of the lumbar drain and my drilling of her skull, looked 'slack' as neurosurgeons say – there was plenty of room for me and Patrik to get underneath it.

'Conditions look lovely,' I shouted to Judith at the other end of the table where she sat in front of a battery of monitors and machines and a cat's cradle of tubes and wires connected to the unconscious Melanie – all the anesthetists can see of the patients are the soles of their feet. Judith, however, had to worry here not just about Melanie's life but about the unborn baby's as well who was being subjected to the same general anaesthetic as his mother.

'Good,' she said.

'Bring the 'scope in and give Patrik a retractor,' I said and, once the heavy microscope had been pushed into position and Patrik was settled in the operating chair, Maria held out a handful of retractors, fanned out like a small pack of cards, from which he took one. I stood at one side, a little nervously looking down the assistant's arm of the microscope.

I told Patrik to place the retractor gently under Melanie's frontal lobe while sucking away the cerebrospinal fluid with a sucker in his other hand. He slowly pulled her brain upwards by a few millimetres.

Look for the lateral third of the sphenoid wing, I told him, and then follow it medially to the anterior clinoid process – these being the important bony landmarks that guide us as we navigate beneath the brain. Patrik cautiously pulled Melanie's brain upwards.

'Is that the right nerve?' asked Patrik.

It most certainly was, I told him, and it looked horribly stretched. We could now see the granular red mass of the tumour over which the right optic nerve – a pale white band a few millimetres in width – was tightly splayed.

'I think I'd better take over now,' I said. 'I'm sorry, but what with the baby and her eyesight being so bad it's not really a training case.'

'Of course,' said Patrik, and he climbed out of the operating chair and I took his place.

I quickly cut into the tumour to the left of the optic nerve and to my relief the tumour was soft and sucked easily – admittedly, most suprasellar tumours do. It did not take long to debulk the tumour with the sucker in my right hand and the diathermy forceps in my left. I gradually eased the hollowed out tumour away from the optic nerves. The tumour was not stuck to the optic nerves and after an hour or so we had a spectacular view of both right and left optic nerves and their junction, known as the chiasm. They look like a pair of miniature white trousers although thin and stretched because of the tumour which I had now removed. On either side were the great carotid arteries that supply most of the blood to the brain and further back the pituitary stalk, the fragile structure that connects the all-important pea-sized pituitary gland to the brain, which co-ordinates all the body's hormonal systems. It sits in a little cavity, known as the sella, just beneath the optic nerves, which is why Melanie's tumour is called a 'supra-sellar' meningioma.

'All out! Let's close up quick and the obstetricians can do the C section,' I announced to the assembled audience. I muttered in an aside to Patrik that I hoped to God that her eyesight would recover.

So Patrik and I closed up Melanie's head and left our colleagues to get on with delivering the baby. As we walked out of the theatre the paediatricians passed us wheeling a paediatric ventilator and resuscitation equipment into the room.

I went off to get a cup of coffee and get some paperwork done in my office. Patrik stayed behind to watch the Caesarean section.

He rang me an hour later. I was sitting at my desk dictating letters.

'It all went fine. She's on the ITU and the baby's next to her.'

'Can she see?' I asked.

'Too early to say,' Patrik said. 'Her pupils are a bit slow ...'

I felt a familiar drag of fear in my stomach. The fact that the pupils of her eyes were not reacting properly to light might just be a temporary anaesthetic effect but it could also mean that the nerves were irreparably damaged and that she was completely blind, even though the operation had seemed to go so well.

'We'll have to wait and see,' I replied.

'The next patient's on the table,' Patrik said. 'Shall we start?'

I left my office to go and join him.

The second patient on the list was a woman in her fifties with a malignant left temporal glioma, a cancerous tumour of the brain itself. I had seen her a week earlier in my outpatient clinic. She had come with her husband, and they held each other's hands as they told me how she had become confused and forgetful over the preceding weeks. I explained to them that her brain scan showed what was undoubtedly a malignant tumour.

'My father died from a malignant brain tumour,' she told me. 'It was terrible to watch him deteriorate and die and I thought that if that happened to me I would not want to be treated.'

'The trouble is,' I said reluctantly, 'it will happen to you anyway. If I treat you, with a bit of luck, you might have some years of reasonable life but if we do nothing you have only a few months left to live.'

In reality this was probably optimistic. The scan showed a foul malignant tumour in her dominant temporal lobe – dominant meaning the half of the brain responsible for speech and language – that was already growing deep into her brain. It was unlikely that she had more than a few months left to live whatever I did, but there is always hope, and there are always a few patients – sadly only a small minority – who are statistical outliers and defy the averages to live for several years.

We had agreed that we should operate. Patrik did most of it, and I assisted him. The operation went well enough, though as soon as Patrik drilled open her head and cut through the meninges, we could see that the tumour was already spreading widely, more widely than in the brain scan done only two weeks earlier. We removed as much of the tumour as we safely could, tangled as it was with the distal branches of the left middle cerebral artery. I did not think we had done her any serious harm, though nor had we done her much good.

'What's her prognosis, boss?' Patrik asked me as he stitched the dura and I cut his stitches with a pair of scissors.

'A few months, probably,' I replied. I told him about her father and what she had said to me.

'It's difficult to do nothing,' I said. 'But death is not always a bad outcome, you know, and a quick death can be better than a slow one.'

Patrik said nothing as he continued to close the woman's meninges with his sutures. Sometimes I discuss with my neurosurgical colleagues what we would do if we – as neurosurgeons and without any illusions about how little treatment achieves – were diagnosed with a malignant brain tumour. I usually say that I hope that I would commit suicide but you never know for certain what you will decide until it happens.

As we stitched her head up I did not expect any problems. Judith took her round on her trolley, pushed by one of the ODAs and nurses, to the ITU while I sat down and wrote an operating note. A few minutes later Judith put her head round the theatre door.

'Henry, she's not waking up and her left pupil is bigger than her right. What do you want to do?'

I swore quietly and quickly walked the short distance to the Intensive Care Unit. In the corner of the room I could see Melanie, and a baby's cot beside her bed, but I hurried past to look at the second patient. With one hand I gently opened her eyelids. The left pupil was large and black, as large as a saucer.

'We'd better scan her,' I said to Patrik who had come hurrying up when he had heard the news. Judith was already re-anaesthetizing the woman and putting a tube down into her lungs so as to put her back on a ventilator. I told Patrik to tell the staff in the scanner that we would be bringing her for a scan immediately and never mind what else they were doing. I wasn't going to wait for a porter. Patrik went to the nurses' desk and picked up the phone while Judith and the nurses disconnected the woman from all the monitoring equipment behind her and with my help wheeled her quickly out of the ITU to take her to the CT scanner. Together with the radiographer we quickly slid her into the machine. I walked back to the control room with its leaded, X-ray-proof window looking out into the room where the patient lay with her head in the scanner.

Impatient and anxious I watched the transverse slices of the scan appear on the computer monitor, gradually working their way up towards where I had been operating. The scan showed a huge haemorrhage deep in her brain, on the side of the operation although slightly separate from it. It was clearly both inoperable and fatal – a post-operative intracerebral haemorrhage, a 'rare but recognized' complication of such surgery. I picked up the phone in the control room and rang her husband.

'I'm afraid I have rather bad news for you ...' I said.

I went round to the surgical sitting room and I lay on the sofa, staring at the sky through the high windows, waiting for her husband and daughter to arrive.

I spoke to them an hour later in the little interview room on the ITU. They collapsed into each other's arms in tears. Dressed in my theatre pyjama suit, I looked on miserably.

As she was going to die the nurses had moved her into a side room where she lay on her own. I took her husband and daughter to see her. They sat down beside her. She was unconscious and mute, her eyes closed, with a lopsided bandage around her head beneath which her bloodied hair hung down. The ventilator which was keeping her alive gently sighed beside her.

'Are you really sure she cannot hear anything we say to her?' her daughter asked me.

I told her that she was in a deep coma but that even if she could hear she would not understand what she heard since the haemorrhage was directly in the speech area of her brain.

'And will she have to stay in hospital? Can't she come home?'

I said that I was certain that she would die within the next twenty-four hours. She would become brain dead and then the ventilator would be switched off.

'She's been taken from us. So suddenly. We were going to do so many things together in the time we had left, weren't we?' her husband said, turning to his daughter as he spoke.

'We weren't ready for this ...' He held his daughter's hand as he talked.

'I trusted you,' he said to me, 'and I still do. Are you certain that she might not wake up? What if she wakes up and finds that we aren't here? It would be so frightening for her although I know she kept on telling us last week that she did not want to be a burden to us.'

'But love is unconditional,' I said and he burst into tears again.

We spoke for a while longer. Eventually I turned to the door saying that I had to leave or I would start crying myself. The husband and daughter laughed at this through their tears. As I left I thought of how I had granted her wish, albeit inadvertently, that she should not die miserably as her father had done.

Back in the operating theatre Patrik was having difficulties stopping the bleeding after removing the disc prolapse in the third and last case on the list. I cursed and abused him half-jokingly and scrubbed up and quickly brought the bleeding under control. We closed up the man's incision together and afterwards I returned to the ITU to see Melanie. She was peacefully asleep and her baby son was asleep in the cot beside her. Her observation chart showed that her pupils were now reacting to light and the nurse looking after her said that all was well. There was a small group of laughing and smiling nurses beside the cot looking at the baby.

Her husband rushed up to me, almost delirious with joy.

'She can see again! You're a miracle worker, Mr Marsh! She woke up from the op and she could see the baby! She said her eyesight's almost back to normal! And our son is fine! How can we ever thank you enough?'

What a day, I thought as I went home, what a day. When

I recounted this story – which I had quite forgotten until then – to the *Holby City* writers gathered round the hotel table, they broke out in little cries of delight and amazement, though whether they used the story about Melanie or not, I do not know.

5

TIC DOULOUREUX

pl.n. brief paroxysms of searing pain felt in the distribution of one or more branches of the trigeminal nerve in the face.

Once I had sawn open the woman's skull and opened the meninges I found to my horror that her brain was obscured by a film of dark, red blood that shouldn't have been there. It probably meant that something had already gone wrong with the operation. The light from the battered old operating lamp above me was so dim that I could scarcely see what I was doing. The possible repercussions for my colleague and me did not bear thinking about. I had to fight to control my mounting panic.

I was operating on a woman with an agonizing facial pain called trigeminal neuralgia (which is also known as *tic douloureux*) – a condition that was considered by her doctors to be inoperable. A television crew was filming the operation for the national news. There were many doctors and nurses, looking down on me like gods through the glass panes of a large dome built into the ceiling above the operating table. Many of the panes in the glass dome were cracked and broken and the view outside through the large windows of the operating theatre was of snow falling onto a grey wasteland of broken machinery and derelict buildings. I often have an audience when operating and I dislike it when things are going badly – but this was many times worse. I had to radiate a calm, surgical self-confidence, which was not what I felt.

This was Ukraine, in 1995. I was 2,000 miles from home, operating without any official permission – probably illegally – doing a dangerous operation on a woman's brain never done in the country before, using second-hand equipment that I had driven out myself from London a few days earlier. My colleague was an obscure junior doctor who had been declared in an interview on the BBC World Service, by the senior professor of neurosurgery in the hospital where I was operating, to be suffering from schizophrenia. Nor was I being paid to do this – indeed, it was costing me a lot of my own money.

I muttered unhappily to myself as I tried to stop my hands shaking: 'Why on earth am I doing this? Is it really necessary?'

I had first gone to Kiev three years earlier in the winter of 1992, almost by accident. I had been a consultant for five years by then and already had a large and busy practice. It was a few months after the collapse of the Soviet Union. An English businessman, hoping to sell medical equipment in Ukraine, rang my hospital to find out if any neurosurgeons were interested in joining him on a trip to Kiev. There was a famous neurosurgical hospital in Kiev and he wanted to take some British neurosurgeons with him to deliver lectures about modern brain surgery and the equipment needed for it. The switchboard operator was rather puzzled by the enquiry and so put the call through to Gail, my secretary, who has the well-earned reputation of being able to solve most problems. I was in my office and she put her head round the door.

'Do you want to go to Ukraine next Thursday?'

'Certainly not. I'm far too busy and I've got a clinic then.'

'Oh go on. You're always saying how interested you are in Russia and you've never been there yet.'

Gail is usually the first person to complain if I cancel an outpatient clinic as she will then have to field all the phone calls from the disappointed and sometimes angry patients and rearrange the appointments, so I had to take her advice seriously.

And so, with two colleagues, I travelled to the newly independent Ukraine. There had never really been a separate Ukrainian state before the Soviet Union fell apart, and it was not at all clear what independence would mean. What was clear was that the country was in utter chaos, with the economy close to collapse. The factories were all closed and everybody seemed to be out of work. The conditions in the hospitals I visited were out of a nightmare.

We had arrived in Kiev early in the morning on the overnight train from Moscow. The line crosses one of the long bridges over the great river Dnieper which flows through Kiev, and as we approached the steep western riverbank we could see the golden domes of the Lavra monastery above us catching the light of the rising sun – a dramatic contrast to the dark railway stations we had passed through during the night and the grim apartment blocks on the outskirts of the city. I had lain in my bunk, under a thin blanket, drifting in and out of sleep, listening to the old-fashioned, rhythmic sound of a train running over bolted rails, travelling southwards across Russia, stopping at dimly lit stations where I could hear incomprehensible announcements echoing over the empty, snow-covered platforms.

It all felt wonderfully strange and yet also strangely familiar – I suppose from the Russian literature in which I had steeped myself in the past. We had only been in Moscow for a few hours. Long enough to stand in Red Square in the dark, in the falling snow, where despite the fall of communism, a huge red flag was still flying, a little listlessly, from the Spassky Tower of the Kremlin. Long enough to have a splendid meal in a hotel that one had to enter through three lines of armed security guards, to find long, shabby corridors with worn, thin carpets and a bewildering number of disconcertingly beautiful girls looking for business. Long enough to understand that with the collapse of the rouble the few hundred dollars we carried in our pockets made us virtual millionaires compared to the impoverished Russians we met.

Once in Kiev we were taken to the Neurosurgical Research Institute, a huge and ugly building with the endless corridors that are the curse of all large hospitals. The corridors were dark and poorly lit. On the walls one could see serious displays of the triumphs of Soviet neurosurgery, with grainy black and white photographs of heroic men in the tall white chef's hats that Soviet surgeons used to wear, interspersed with hammers and sickles, red stars, inspirational slogans and photographs of scenes from the Great Patriotic War, as the Russians call the Second World War. But everything, from the building itself to the pictures on the walls, and the stale air which smelt of cheap tobacco and some odd, sickly smelling disinfectant, felt tired and faded. We were ushered into the office of Academician Romadanov, an old, imposing and very eminent man and the director of the institute. He was tall, with a large head and a mane of white hair, and he wore a high-collared white coat, buttoned round his throat. He looked, however, as tired and faded as the corridors, and was in fact to die one year later. After the usual introductions – all conducted through an interpreter – we sat down round the long table in his office.

'Why have you come here?' he asked angrily. 'As tourists? To amuse yourselves by seeing all our problems? This is a very difficult time for us.'

We tried to answer diplomatically and talked of friendship

and professional collaboration and international cooperation. He looked unconvinced and he was, of course, entirely right.

We were then shown round the famous institute by one of his assistants.

'This is the largest neurosurgical hospital in the world,' we were told.

'There are eight departments and five floors and four hundred beds.'

I was amazed – my own hospital, one of the largest neurosurgical units in Britain, had only fifty beds. We traipsed up and down the stairs and along corridors and visited each identical department in turn.

We started on the ground floor.

'This is the Department of Posterior Fossa Tumours,' we were told.

As we came through the doors the staff came out to meet us, to shake hands and to be photographed with us. I was told all about the wide range of operations that went on in the department, although any detailed questions on my part were usually met with rather vague answers. We went through exactly the same ritual in the seven other departments. When I asked if we could see the operating theatres I was told that they were being redecorated and were closed. We saw scarcely a single patient.

We delivered our lectures. The few questions afterwards showed a complete and utter lack of understanding of what we had been trying to explain. We returned to our hotel. As with the hotel in Moscow, there were beautiful young women to be seen everywhere. I was told that they were not professional prostitutes but respectable women desperately trying to make some money. One session with a western businessman was, at that time, worth more than a whole month's income. Embarrassed and fascinated, we made our way shyly past them and retreated to one of our rooms to drink duty-free whisky, confused and shocked by the surreal discrepancy between what we had seen and what we had been told as we toured the hospital.

The next day I was taken to the Emergency Hospital on the eastern side of the city. I had asked to see how trauma was managed and my guides – a little reluctantly – had agreed to take me. We arrived in the late afternoon. The light was starting to fade. The hospital was ten storeys high, apparently with eight hundred beds. It was only ten years old but already looked derelict. We approached it through a wasteland of broken buildings and those gigantic, incomprehensible pipes that always seem to surround Soviet buildings, on which pure white snow was starting to fall from a leaden sky. At one side there was a large and ramshackle open-air market, with battered zinc-covered huts displaying rather sad little collections of cheap cosmetics and vodka. Decrepit Lada and Moskvitch and Volga cars were parked in utter disorder. Everything was grey, colourless and drab in the way that only Soviet cities could be. Collecting the illegal rent paid by the market traders was, I subsequently heard, an important part of the hospital director's job and a useful source of income for the officials of the city's health administration.

The electricity supply had failed and much of the hospital was in pitch darkness. The whole place stank of ammonia – the hospital had run out of disinfectants and only ammonia was available for cleaning. The building seemed almost uninhabited. I was taken to one of the dark operating theatres – a huge cavernous place with a large window looking out onto what appeared to be a bombsite. Flurries of snow could be seen there, caught in the dim light from the window of the theatre. An operation was going on. A surgeon was 'operating' on a paralysed man, paralysed from the neck down in an accident some years previously, I was told. There was a small tray of battered instruments beside him that looked as though they came from a scrapyard. The patient was lying on his side and was partially covered with old curtains with a faded floral pattern. The surgeon had inserted several large needles into his spine and was injecting cold saline through them into the spinal canal. This was, apparently, supposed to stimulate the spinal cord to recover. The reflex movements in the paralysed man's legs that the injections produced were greeted with cries of excitement and seen as evidence that the treatment was working.

While I was walking along one particularly dark and dismal corridor, a young man came hurrying up to me like an enthusiastic spaniel. It was the surgeon I had seen 'operating' on the paralysed man.

'This is neurosurgical department,' he announced in broken English. 'There are three departments of emergency neurosurgery. I am Igor Kurilets, director of spinal emergency department.' I expected the long and tedious description to continue. I was quickly becoming familiar with the litany of departments and beds and achievements with which one was greeted when one visited a Ukrainian hospital and expected to be reassured that Ukrainian emergency spinal neurosurgery was the equal of the world, if not better.

'Everything terrible here!' he said.

I liked Igor immediately. Apart from Academician Romadanov, he was the only doctor I met on my first visit who seemed able to admit openly that the medical situation in Ukraine – at least in neurosurgery – was dire. The Soviet Union had excelled at producing guns and rockets but failed miserably at producing decent health care. Although there were impressive-sounding research institutes and thousands of professors, the reality was of poorly trained doctors and poorly equipped hospitals that were often little better than what one might find in the Third World. The Soviet Union, it used to be said, 'is Upper Volta with rockets' – Upper

Volta, as it was then called, being the poorest country in Africa. Most of the doctors I met, driven by a mixture of shame, patriotism, envy and embarrassment, felt compelled to deny this, and did not welcome people like Igor who dared to point out the emperor's lack of clothes. Soviet culture had never encouraged criticism and had gone to great lengths to isolate its citizens from the rest of the world. Despite the fall of the Soviet Union, newly independent Ukraine still had the same leaders as in the past, but the country and its people were suddenly exposed to the outside world and the huge gulf that had developed between western and Eastern European medicine.

Before I left Kiev after this first visit I attended a meeting at the Ministry of Health. An expressionless, florid-faced bureaucrat, the umpteenth secretary of some umpteenth department of this or that, walked round the long table handing out his business card, where his umpteen titles were all duly listed. The more important bureaucrats, I had noticed, would have so many titles and appointments that more than one card was needed to list them all. This man was only a one-card bureaucrat so clearly not too important.

I soon lost interest in what was being said. Besides, it all had to be slowly translated, which made it doubly tedious. The room, panelled in cheap plywood like most Soviet government offices, had tall windows that looked out onto an attractive park. Snow was starting to fall again. A police van was disgorging heavily armed riot police in grey uniforms with German shepherd dogs. Both dogs and men seemed to be jumping out of the back of the vehicle with great enthusiasm. We had seen a demonstration by the Ukrainian Nationalist Party going on outside the nearby Parliament building on our way to the Health Ministry so perhaps the policemen and their dogs were looking forward to a good fight. The English businessman who had brought me to Ukraine was sitting next to me and leant over to whisper that the riot police were the pimps for the girls we had seen in the hotel.

There was an inconsequential and meaningless conversation about furthering international medical cooperation. At the end I commented that I would be happy to arrange for a Ukrainian neurosurgeon to come to London to work with me but added that it could only be one man, Dr Kurilets, the director of the obscure and unimportant spinal trauma department in the Emergency Hospital. This was a post he had been given, he later told me, as a form of demotion, Soviet medicine having little interest in the lame or paralysed. I knew well enough that it was highly unlikely that Igor, low down the ladder as he was, would be allowed out of the country but it seemed worth a try and I was damned if I was going to invite one of the elderly, dissembling professors. The bureaucrat looked nonplussed and I returned to London via Moscow that afternoon.

Nine months later, I had almost forgotten the high hopes which I had had when leaving Kiev, when, quite unexpectedly, a Christmas card from Igor arrived, enclosing a letter from Academician Romadanov asking me to bring Igor to London and show him modern neurosurgery.

What had begun on my part as casual tourism became more serious when Igor started to encounter opposition from the Ukrainian medical establishment. After three months working with me in London he had returned to find that his patron Academician Romadanov had died. Rather than find a new source of patronage and support (an essential requirement in Ukrainian society and known as 'a roof over your head') Igor proceeded to declare in public that Ukrainian neurosurgery was primitive and backward and that a revolution was required. Matters were not helped by the fact that a somewhat Byzantine struggle for succession to

Academician Romadanov's post was going on. The position came with important perks such as a large apartment and a chauffeur-driven car. Igor's own boss had been hoping to get the job and his chances were not helped by Igor's insubordination.

The next few years were very difficult indeed for Igor as he struggled to reorganize and modernize his department along western lines. There was a long series of official denunciations, investigations and threatening phone calls. For a time he slept in a different room each night. How he coped with all this I cannot even begin to imagine.

I realized that my rather naive wish to help him had caused as many problems as it had solved and yet I could not very well abandon Igor. So each time that his 'detractors', as he put it, tried to 'bump' him – to close his department or sack his staff – I would do what I could to help, although admittedly this was usually from a long, safe distance. And when I did go out to Kiev I knew that I could always escape back home again, however unpleasant some of my encounters with senior bureaucrats might be. With Igor's help I wrote articles in Ukrainian newspapers and staged press conferences. I drove second-hand medical equipment out to Kiev and brought his own junior doctors over to London to work with me. I carried out brain operations that had not previously been performed in Ukraine. In retrospect, given the poor operating conditions and the implacable hostility of the medical establishment, what I did in those years seems to me now to have been verging on lunacy. It certainly required a self-confidence and independence that I was subsequently to lose.

Despite the inauspicious start and my ignominious panic the operation on the woman with trigeminal neuralgia was a great success and she appeared on the national TV news next day to say that she was pain-free for the first time in many years. I flew back to Poland to collect my car, which I had left with a friend. I had driven the microscope which I had used for the operation to his home in western Poland and Igor had then come from Ukraine in an old van to collect me and the equipment.

On our way to the airport we made a detour to the Bessarabian Market in central Kiev. The Bessarabian Market is Kiev's equivalent of Les Halles or Covent Garden – a large circular nineteenth-century building with a ribbed cast-iron glass roof. Below is a market, with fierce but friendly women in brightly coloured headscarves, standing behind pyramids of beautifully displayed fruit and vegetables and pickles. There is a flower section – the Ukrainians give flowers to each other on any social occasion – and a meat section, with whole hog's heads and mounds of fresh meat, and the rear-quarters of pigs, hanging up from hooks like pairs of trousers. There is a directness and rawness, a rough beauty to the place that is typical of Ukraine but which is starting to disappear now that supermarkets have arrived. Igor was later to tell me that the Bessarabian Market was still functioning only because it had become something of a tourist attraction. He suddenly became quite excited and pointed to one of the fish stalls.

'Very rare!' he said, pointing at three long smoked eels in a glass cabinet. He bought one of them and gave it to me as a present. It smelt rather awful.

'Very unusual!' he said proudly. 'They are in Red Book!'

'What's the Red Book?' I asked.

'Book of animals soon dead. None left. You are lucky to have one,' he said happily.

'But Igor, this could be the last Ukrainian eel!' I said, looking at the long and once-beautiful creature, who had been swimming, glittering, in some remote Ukrainian river and was now smoked and dead and wrapped in a grubby Giorgio Armani plastic bag. I took it from Igor and dutifully packed it in my suitcase.

On my return to London a few days later I threw the smoked eel into my back garden, since I could not face eating it and I thought an itinerant fox, who I often see walking quietly past early in the morning, might like it. The eel had disappeared the next day but I was rather saddened to find it later a few yards away under a bush – it had been rejected even by the fox. So I dug a hole and buried it, the last Ukrainian eel, in an overgrown flowerbed at the end of the garden.

6

ANGOR ANIMI

n. the sense of being in the act of dying, differing from the
fear of death or the desire for death.

Just as I had first gone to Ukraine out of curiosity and not
from any particular wish to help Ukrainians – although I
have now been working there for more than twenty years – I
had become a doctor not from any deep sense of vocation
but because of a crisis in my life.

Until the age of twenty-one I had followed the path that
seemed clearly laid out for me by my family and education. It was a time when people from my background could
simply assume that a job was waiting for them – the only
question was to decide what you wanted to do. I had received a private and privileged English education in a famous
school, with many years devoted to Latin and Greek, and
then to English and History. I took two years off on leaving
school, and after several months editing medieval customs
documents in the Public Record Office (a job organized by
my father through his many connections), spent a year as a
volunteer teaching English literature in a remote corner of
West Africa. I then went up to Oxford to read Politics, Philosophy and Economics.

I was destined, I suppose, for an academic or adminis-
trative career of some kind. During all these years I had

received virtually no scientific education. Apart from a maternal great-grandfather who had been a village doctor in rural Prussia in the early decades of the last century there was nothing medical or scientific in my family background. My father was an eminent English human rights lawyer and academic, and my mother a German refugee from Nazi Germany who would probably have been a philologist if she hadn't refused to join the women's branch of the Hitler Youth – the League of German Maidens – and hence had been denied entry into university. Apart from the one doctor in Prussia my ancestors on both sides of my family were teachers and clergymen and tradesmen (although my uncle had been a Messerschmitt fighter pilot in a crack squadron until shot down in 1940).

While at Oxford, I fell in love, love which was unrequited, and driven by self-pitying despair I had, to my father's deep dismay, abandoned university and run away to work as a hospital porter in a mining town in the north of England, trying to emulate Jack Nicholson heading off for Alaska at the end of the movie *Five Easy Pieces*. I spent six months there, spending the days lifting patients on and off operating tables, cleaning walls and equipment and assisting the anaesthetists.

I lived in a small room in a semi-derelict old fever hospital with a corrugated iron roof on the muddy banks of the polluted River Wansbeck. It was a few miles away from the coast where the beaches were black with sea-coal. There was a huge coal-fired power-station which I could see in the distance from my room, with high chimneys pouring white smoke and steam into the wind off the sea. At night the rising steam was lit by the arc-lamps that stood over the mountains of coal beside the turbine halls, over which I could see bulldozers crawling beneath the stars. I wrote second-rate, self-obsessed poetry in which I described this view as being of both heaven and hell. Full of youthful melodrama I saw myself as living in a world as red as blood and as white as snow – although the surgery I saw was not especially bloody and the winter was mild, without any snow.

I was profoundly lonely. In retrospect I was obviously trying to realize my own unhappiness by working in a hospital, in a place of illness and suffering, and perhaps I was curing myself of my adolescent angst and unrequited love in the process. It was also a ritual rebellion against my poor, well-meaning father who up till then had largely determined the course of my life. After six months of this I desperately wanted to come home – both to my family but also to a professional middle-class career, although one of my own choosing. Having spent six months watching surgeons operating I decided that this was what I should do. I found its controlled and altruistic violence deeply appealing. It seemed to involve excitement and job security, a combination of manual and mental skills, and power and social status as well. Nevertheless, it was not until eight years later when as a junior doctor I saw that first aneurysm operation that I discovered my vocation.

I was fortunate that my college at Oxford allowed me to come back after my year away to complete my degree and I was later accepted to study medicine at the only medical school in London which took students without any scientific qualifications. Having been rejected by all the other London Medical Schools since I had neither O-levels nor A-levels in science I had telephoned the Royal Free Medical School. They asked me to come for an interview next day.

The interview was with an elderly, pipe-smoking Scot, the Medical School Registrar, in a small and cramped office. He was to retire a few weeks later and perhaps he let me in to the Medical School as a kind of joke, or celebration, or perhaps his mind was elsewhere. He asked me if I enjoyed fly-fishing. I replied that I did not. He said that it was best to see medicine as a form of craft, neither art nor science – an opinion with which I came to agree in later years. The interview took five minutes and he offered me a place in the Medical School starting three weeks later.

Selection for medical schools has become a rather more rigorous process since then. I believe the Medical School at the huge London hospital where I now work uses role-playing with actors, along with many other procedures, to select the doctors of the future. The nervous candidates must show their ability to break bad news by telling an actor that their cat has just been run over by a car. Failure to take the scenario seriously, I am told, results in immediate rejection. Whether this is any better than the process I went through remains, I believe, unproven. Apparently the actors help select the successful candidates.

I joined what was called a First MB course – which was a year's crash course in basic science, and which led on to the Second MB course, which was the standard five-year course of undergraduate medical training. This was the last year that the medical school ran a First MB Course and the department was something of a scientific and academic backwater, with the course taught by a number of eccentric and often embittered scientists – although many of them were at the very beginning of their careers, and rapidly moved on elsewhere. One became a famous science writer, another eventually became a peer and Chairman of the Tory party. The others were older teachers approaching retirement, some of whom did not bother to hide their dislike of the slightly odd mix of First MB students – a stockbroker, a Saudi princess, a Ford truck salesman – mixed up with other, younger students who had poor A-level results (and one, it turned out, who had forged them). Our days were spent gradually dissecting and dis-assembling large white rabbits for biology, titrating chemicals for chemistry and failing to understand the physics lectures. Some of the lecturers were inspiring, some risible. The atmosphere was anxious, verging on hysterical – we were all desperate to become doctors and most of us felt a failure for some reason or other, although as far as I can remember we all passed the final examination.

I then spent two years of pre-clinical studies in the medical school – anatomy, physiology, biochemistry and pharmacology – followed by three years as a clinical student in the hospital. The anatomy involved the students being divided up into small groups and each group was given an embalmed cadaver which we slowly took apart over the course of the year. Not especially attractive to begin with, the cadavers were a sorry sight by the year's end. The bodies were kept in the Long Room – a large and high attic space with skylights, with half a dozen trolleys on either side with sinister shapes covered by green tarpaulins. The place smelled strongly of formaldehyde.

On the first day of the course, holding our newly purchased dissection manuals with a few instruments in a small canvas roll, we queued up a little nervously on the stairs leading to the Long Room. The doors were opened with a flourish by the Long Room attendant and we went in to be presented to our respective, intact corpses. It was a traditional part of medical education going back hundreds of years but has now been largely abandoned. As a surgeon one has to learn real anatomy all over again – the anatomy of a living, bleeding body is quite different from the greasy, grey flesh of cadavers embalmed for dissection. The anatomy we learnt from dissection was perhaps of limited value, but it was an important initiation rite, marking our transition from the lay world to the world of disease and death and perhaps inuring us to it. It was also quite a sociable process as one sat with a group of fellow students around one's cadaver, picking and scratching away at dead tissue, learning the hundreds of names that had to be learnt – of the veins and arteries and bones and organ parts and their relations. I remember being particularly fascinated by the anatomy of the hand. There was a plastic bag of severed hands in the anatomy department in various stages of dissection from which I liked to make elaborate, coloured drawings, in imitation of Vesalius.

In 1979 I emerged onto the wards of the hospital where I had trained wearing the long white coat of a junior doctor as opposed to the short white coat of a medical student. I felt very important. Other hospitals, I later noticed to my confusion, had the medical students in long white coats and the junior doctors in short ones. Like a badge of office I proudly carried a pager – known colloquially as a bleep – in the breast pocket with a stethoscope, a tourniquet for blood-taking and a drug formulary in the side pockets. Once you had qualified from medical school you spent a year as a junior house officer – a sort of general dogsbody – with six months working in surgery and six months in medicine. If you wanted a career in hospital medicine as a surgeon or physician – as opposed to becoming a GP – you tried to get a housejob in the teaching hospital where you had trained as a student, so as to make yourself known to the senior doctors, on whose patronage your career entirely depended.

I wanted to be a surgeon – at least I thought I did – so I managed to get a job on a surgical 'firm', as it was called, in my teaching hospital. The firm consisted of a consultant, a senior registrar and a junior registrar and the houseman. I worked '1 in 2', which meant I did a normal working day five days a week, but also was on call every other night and every other weekend, so I was in the hospital for about 120 hours a week. My predecessor had handed me over the bleep with a few words of advice about how to keep the boss happy and how to help patients who were dying – neither subject being dealt with in the lectures and textbooks. I enjoyed the feeling of power and importance the long hours gave me. In reality, I had little responsibility. The days and nights were spent clerking in patients, taking blood, filling up forms and chasing up missing X-rays. I usually got just enough sleep, and I became used to being disturbed at night. Occasionally I assisted in theatre, which meant long hours standing still, holding patients' abdomens open with retractors while my seniors rummaged about. Looking back now, thirty years later, my sense of my own importance at that time seems quite laughable.

Much as I liked being part of the small army of junior doctors in the hospital, as the months as a surgical house-man passed I became increasingly uncertain as to what I was going to do with my medical career. The reality of surgery had proved quite different from my superficial impressions of it when a theatre porter. Surgery seemed to involve unpleasant, smelly body parts, sphincters and bodily fluids that I found almost as unattractive as some of the surgeons dealing with them, although there were a few surgical teachers in the hospital without whose influence I would never have become a surgeon. It was their kindness to patients, as much as their technical skill, which I found inspiring. I saw no neurosurgery as a medical student or houseman. The neurosurgical operating theatre was out of bounds, and people spoke of it with awe, almost alarm.

My next six months as a houseman were in a dilapidated old hospital in south London. The building had housed a workhouse in the nineteenth century and it was said it had not yet escaped its dismal previous reputation with the local population. It was the sort of hospital which made the British public's devotion to the NHS quite incomprehensible, with the patients housed like cattle in the old workhouse rooms – large and ugly rooms with dozens of beds lined up on either side. The Casualty department was on the ground floor and the Intensive Care Unit on the first floor immediately above it, but there was only one lift in the hospital which was one quarter of a mile away down the main hospital corridor. If a patient had to be transferred urgently from Casualty to the ITU it was the task of the junior houseman on call, with the help of a porter, to push the patient's bed all the way down the corridor from one end of the hospital to the other, take the lift, and then push the patient and bed all the way back again. I tried to do this as quickly as possible, pushing people in the corridor out of the way and commandeering the large and rattling old lift, creating a sense of drama and urgency. I doubt if it was clinically necessary but it was how it was done on TV and was good fun. Even though I got little sleep at night there was a doctors' mess and bar run by a friendly Spanish lady who would cook me a meal at any time of night. There was even a lawn outside the main building where I would play croquet with my fellow junior doctors when we had the time.

It was busy work with more responsibility than my first job as a house surgeon, and with much less supervision. I learnt a lot of practical medicine very quickly but they were not always enjoyable lessons. I was at the bottom of a little hierarchy in the 'firm'. My job was to see all the patients – most of whom were admitted as emergencies through the Casualty department – when they arrived and to look after the ones already on the wards. I learnt very quickly that I did not ring up my seniors about a patient without having first seen the patient myself. I had done this on my first night on call, asking my registrar's advice in advance of going to see a patient the nurses had called me about, and received a torrent of abuse in reply. So, anxious and inexperienced, I would see all the patients, try to decide what to do, and only dare to ring my seniors up if I was really very uncertain indeed.

One night, shortly after I had started, I was called in the early hours to see a middle-aged man on the ward who had become breathless, a common enough problem on a busy emergency medical ward. I got out of bed and pulled on my white coat (I slept with my clothes on since one rarely got more than an hour or two of sleep without being called to Casualty or the wards). I walked onto the long and darkened Nightingale ward with its twenty beds on either side facing each other. Restless, snoring, shifting shapes lay in them. Two nurses sat at a desk in the middle of the room, a little pool of light in the darkness, doing paperwork. They pointed to the patient they wanted me to see.

'He came in yesterday with a query MI,' one of them said, 'MI' being short for a myocardial infarct, or heart attack.

The man was sitting upright in his bed. He looked terrified. His pulse was fast and he was breathing quickly. I put my stethoscope to his chest and listened to his heart and breath sounds. I ran an ECG – an electro-cardiogram which shows the heart's rhythm. It seemed normal enough to me so I reassured him and told him that there was nothing seriously wrong with his heart

'There's something the matter Doc,' he said, 'I know there is.'

'Everything's all right, you're just anxious,' I said a little impatiently, keen to get back to bed. He looked despairingly at me as I turned away. I can still hear his laboured breathing now, the sound following me like an accusation, as I walked away between the rows of beds with their huddled, restless shapes. I can still hear the way in which, as I reached the doors to the ward, his breathing abruptly stopped, and the ward was suddenly silent. I raced back to the bed, panic-struck, to find him slumped in his bed.

'Put out a crash call!' I shouted to the nurses as I started to pound his chest. After a few minutes my colleagues tumbled bleary-eyed onto the ward and we spent half an hour failing to get his heart going again. My registrar looked at the earlier ECG trace.

'Looks like there were runs of V-Tach,' he said disapprovingly. 'Didn't you notice that? You should have rung me.' I said nothing in reply.

It used to be called *angor animi* – the anguish of the soul – the feeling that some people have, when they are having a heart attack, that they are about to die. Even now, more than thirty years later, I can see very clearly the dying man's despairing expression as he looked at me as I turned away.

There was a slightly grim, exhilarating intensity to the work and I quickly lost the simple altruism I had had as a medical student. It had been easy then to feel sympathy for patients because I was not responsible for what happened to them. But with responsibility comes fear of failure, and patients become a source of anxiety and stress as well as occasional pride in success. I dealt with death on a daily basis, often in the form of attempted resuscitation and sometimes with patients bleeding to death from internal haemorrhage. The reality of cardio-pulmonary resuscitation is very different from what is shown on TV. Most attempts are miserable, violent affairs, and can involve breaking the ribs of elderly patients who would be better left to die in peace.

So I became hardened in the way that doctors have to become hardened and came to see patients as an entirely separate race from all-important, invulnerable young doctors like myself. Now that I am reaching the end of my career this detachment has started to fade. I am less frightened by failure – I have come to accept it and feel less threatened by it and hopefully have learned from the mistakes I made in the past. I can dare to be a little less detached. Besides, with advancing age I can no longer deny that I am made of the same flesh and blood as my patients and that I am equally vulnerable. So I now feel a deeper pity for them than in the past – I know that I too, sooner or later, will be stuck like them in a bed in a crowded hospital bay, fearing for my life.

After finishing my year as a houseman I returned to my teaching hospital in north London to work as a senior house officer on the Intensive Care Unit. I had decided, with diminishing conviction, to try to train as a surgeon and working in intensive care was seen as a useful first step. The job mainly involved filling in forms, putting up drips, taking blood and occasionally more exciting invasive procedures, as they are called, such as inserting chest drains or IVs into the large veins of the neck. All the practical instruction was given by the more experienced junior doctors. It was while working on the ITU that I had gone down to the operating theatres and seen the aneurysm operation that prompted my surgical epiphany.

Now that I knew exactly what I wanted to do my life became much easier. A few days later I went to find the neurosurgeon I had watched clipping the aneurysm and told him I wanted to become a neurosurgeon. He told me to apply for the neurosurgical SHO job in his department which was shortly to be advertised. I also spoke to one of the senior general surgeons on whose firm I had been as a student. An exceptionally kind man – the kind of surgical teacher I came close to worshipping – he immediately arranged for me to go and see two of the most senior neurosurgeons in the country both to make myself known as a would-be brain surgeon and to plan my career. Neurosurgery was a small world in those years, with fewer than a hundred consultants in all of the UK. One of the senior surgeons I went to see was at the Royal London in the East End. A very affable man, I found

him in his office smoking a cigar. The walls were lined with
photographs of Formula 1 racing cars since he was, I learned,
the doctor for Formula 1 racing. I told him of my deep desire
to be a neurosurgeon.

'What does your wife think about it?' was his first question.

'I think she thinks it's a good idea, Sir,' I said.

'Well, my first wife couldn't stand the life so I changed her
for a different model,' he replied. 'It's a hard life, you know,
training in neurosurgery.'

A few weeks later I drove down to Southampton to see another senior neurosurgeon. He was equally friendly. Balding,
with red hair and a moustache, he looked more like a jovial
farmer than what I expected a neurosurgeon to look like. He
sat at a desk covered in piles of patients' notes that almost
hid him from view. I told him about my ambition to become
a neurosurgeon.

'What does your wife think about it?' he asked. I assured
him that all would be well. He said nothing for a while.

'The operating is the easy part, you know,' he said. 'By
my age you realize that the difficulties are all to do with the
decision-making.'

7

MENINGIOMA

n. a benign tumour arising from the fibrous covering of the brain and spinal cord; usu. slow-growing, produces symptoms by pressure on the underlying nervous tissue.

On Monday morning I had awoken at seven, to the sound of heavy rain. It was February and the sky, seen dimly through my bedroom windows, was the colour of lead. There was a long operating list ahead of me but I doubted if I would be able to finish it since I knew that the hospital was overflowing once again, and that there was a shortage of beds. The day would end with the misery of my having to apologize to at least one patient, who would have been kept waiting all day, nil-by-mouth, starved and anxious, on the off-chance that a post-operative bed might become available, to be told that their operation would have to be postponed.

So, cursing the weather, with the wind and rain against me, and cursing the hospital's bed state, I cycled to work. I was late for the morning meeting and sat down next to one of my colleagues, a neuroradiologist whose interpretation of brain scans – a very difficult skill – is second to none and on whose advice I depend to save me from making mistakes. I asked Anthony, the registrar who had been on call for emergency admissions overnight to present the admissions. He was sitting at the computer at the front of the room and had

86

been waiting for me to arrive. Anthony was quite junior and tended to be a bit gung-ho, not an unusual characteristic in a surgeon, but one which most neurosurgeons lose as they become more experienced.

'There was nothing very interesting last night,' he replied.

I looked at him and irritably told him that the simple, everyday problems were often the most important ones.

He looked hurt by my criticism, and I momentarily regretted my bad manners.

'This is a ninety-six-year-old woman who has been living independently and has been starting to fall at home,' he said. 'She's got severe aortic stenosis – you can hear the cardiac murmur from the end of the bed. She's got a left hemiparesis and can't walk but is fully orientated.'

I asked one of the most junior doctors sitting in the front row for the most likely diagnosis.

'The only condition we might treat in somebody that age would be a chronic subdural,' he replied confidently.

I asked him about the significance of the aortic stenosis.

'It means that a general anaesthetic would probably kill her.'

I told Anthony to show us the scan. He turned to the computer keyboard. He typed in a number of passwords but it took several minutes before the website linking us to the local hospitals, from where most of our patients come, appeared. While he fiddled with the computer the other junior doctors laughed and joked about the hospital's IT systems, while trying to help him locate the patient's scans.

'The software for transferring scans is complete crap ... Try refreshing, Anthony – no, go to View, then tile – doesn't seem to be working. Drag it over to the left. It's useless. Try going back to login ...' Eventually the old woman's brain scan suddenly flashed up on the wall in front of us. It showed a thick layer of fluid between the inside of her skull and the surface of her brain, distorting the right cerebral hemisphere.

It was yet another old person with a chronic subdural – the commonest emergency in neurosurgery. The rest of the brain didn't look too bad for her age, and much less shrunken than for most ninety-six-year-olds.

'My father died at that age with Alzheimer's,' I said to the trainees. 'His brain on his brain scan looked like a Swiss cheese plant, there was so little left.'

'So, Anthony,' I continued, 'what's the problem?'

'It's an ethical problem. She says she would rather die than have to leave her home and end up in a nursing home.'

'Well, that's not unreasonable. Have you ever worked on a psycho-geriatric ward or nursing home?'

'No,' he replied.

I started to talk about how I had once worked as a psycho-geriatric nursing assistant. Looking after a ward of twenty-six doubly incontinent old men was not easy. As the population gets older and older there are going to be more and more scandals in the media about abuse in old people's homes. By 2050 a third of the population of Europe will be over sixty. My first boss in general surgery – a lovely man – ended his days in a nursing home because of dementia. His daughter told me he kept on saying that he wanted to die but he was terribly fit and it took ages. He used to have a cold bath every morning when he was younger.

'Well we can't just let her die,' one of the registrars in the back row said, interrupting my monologue.

'Why not?' I said. 'If that's what she wants.'

'But she might be depressed. She might change her mind.'

We discussed this for a while. I pointed out that his comments applied well enough to younger people who had many years of life ahead of them if they didn't commit suicide, but I was uncertain if it applied to somebody aged ninety-six who had little chance of getting home again.

I asked Anthony what he thought were the chances of her returning to an independent life in her own home if we operated.

'Not very good, at her age,' he replied. 'Though I suppose she might get back for a while but she'll still end up in a nursing home sooner or later if the aortic stenosis doesn't finish her off first.'

'So what should we do?' I said to the room. There was an uncomfortable silence. I waited for a while.

'The only next of kin is a niece. She's coming in this morning,' Anthony told us.

'Well, any decisions will have to wait until then.'

My radiological colleague leant over towards me and spoke quietly.

'I always find these cases by far the most interesting,' he said. 'The young ones,' he nodded towards the row of junior doctors, 'all want to operate, and want big, exciting cases – that's fair enough at their age, but the discussions about these everyday cases are fascinating.'

'Well, I was just like that once upon a time,' I replied.

'What do you think will happen to her?' he asked.

'I don't know. She's not my case.' I turned to the assembled doctors. 'There's ten minutes left,' I said. 'Shall we look at one of the cases on my list for today?' I gave the patient's name to Anthony and he summoned up a brain scan onto the wall, with greater success than with the first case. The scan showed a huge tumour – a benign meningioma – pressing on the left side of the patient's brain.

'She's eighty-five years old,' I began. 'When I went in to neurosurgery thirty-two years ago, when you lot were probably still in trainer pants, we just didn't operate on people this old. Anybody over the age of seventy was simply considered too old. Now there doesn't seem to be any age limit.' So I gave them the history, the story about the patient.

*

I had first seen Mrs Seagrave several weeks earlier in the outpatient clinic. She was the highly articulate widow of an eminent doctor and she came accompanied by her three very professional and equally articulate middle-aged children – two daughters and a son. I had to go to another room to bring in some extra chairs. The patient, a short and dominating woman with long grey hair, well turned out and looking younger than her age, marched into the room in an authoritative way. She sat in the chair beside my desk and her three children sat in a row facing me, a polite but determined chorus. As with most people with problems affecting the front of the brain she had little insight, if any, into her difficulties.

Having introduced myself I asked her, with the wary sympathy of the doctor anxious to help but anxious also to avoid the emotional demands patients make of their doctors, to tell me about the problems that had led to her having a brain scan.

'I'm perfectly all right!' she declared in a ringing tone. 'My husband was the professor of gynaecology at St Anne's. Did you know him?'

I said that I did not – that he had been rather before my time.

'But it's just outrageous that they,' she gestured towards her children sitting opposite her, 'won't let me drive. I really can't manage without a car. It's all so sexist ... If I was a man they'd let me drive.'

'But you are eighty-five years old ...' I said.

'That has nothing to do with it!'

'And there is the matter of the brain tumour,' I added, pointing to the monitor on my desk. 'Have you seen your brain scan before?'

'No,' she said. 'Well, isn't that interesting.' She looked

thoughtfully at the scan for a while, showing the large mass, the size of a grapefruit, compressing her brain. 'But I really must be allowed to drive. I can't manage without it.'

'If you'll excuse me,' I said, 'I would like to ask your children a few questions.'

I asked about the difficulties their mother had had in recent months. I think they were a little reticent about drawing attention to her problems in her presence – and she constantly interrupted, disputing what they said and, above all, complaining about the fact they would not let her drive. Between the three of them they gave me to understand that their mother had become confused and forgetful. At first, naturally enough, they had attributed this to her age but her memory became steadily worse and she was seen by a geriatrician who had arranged a brain scan. Brain tumours like hers are a rare but recognized cause of dementia and can be surprisingly large by the time they start to cause problems. It was always possible, however, that she was suffering from Alzheimer's disease as well as having a brain tumour and an operation to remove the tumour, I therefore told them, was not guaranteed to make her better. It also came with a very real danger of making her much worse. The only sure way of knowing whether the tumour was responsible for her problems, however, would be by removing it. The problem, I told them, was that it was impossible to predict from the scan just how great the risk was of making her worse. It is a question of how stuck the surface of the tumour is to the surface of the brain and until you operate you cannot tell how easy or difficult it will be to separate the tumour away from the underlying brain. If it is stuck, the brain will be damaged and she could be left paralysed down the right side of her body and unable to communicate, as each half of the brain controls the opposite side of the body and the function of speech is on the left side of the brain.

'Can't you just remove part of the tumour,' one of the daughters asked, 'and leave the part stuck to the brain?'

I explained that this rarely worked since these tumours are often quite solid and if you leave a rigid shell of tumour behind, the brain remains compressed and the patient doesn't get any better. And the tumour can grow back again.

'Well, how often is the tumour stuck to the brain?' the other daughter asked.

'Well, it's a bit of a guess, but I suppose twenty per cent.'

'So there's a one in five chance of making her worse?'

In fact it was probably slightly more than that because every time you open somebody's head there's a one to two per cent risk of catastrophic haemorrhage or infection, and that risk is probably slightly greater in somebody her age. The only certainty was that she would slowly get worse if we did nothing – but, I added hesitantly, hoping that Mrs Seagrave herself would not notice, one might argue that given her age it was best simply not to operate and accept that she would slowly deteriorate before she died.

One of the daughters asked if any treatment other than surgery might help. With Mrs Seagrave continuing to interrupt, complaining about the monstrous injustice of her not being allowed to drive, I explained that radiotherapy and chemotherapy were of no use with tumours of this type. It was fairly obvious that their mother was not capable of following the conversation.

'What would you do if it were your mother?' the son asked.

I hesitated before I answered because I was not sure of the answer. It is, of course, the question that all patients should ask their doctors, but one most are reluctant to ask since the question suggests that doctors might choose differently for themselves compared to what they recommend for their patients.

I replied slowly that I would try to persuade her to have
the operation if we – and I gestured to the four of them
as I spoke – all felt she was losing her independence and
heading for some kind of institutional care. But I said that
it was very difficult – that it was all about uncertainty and
luck. I was sitting with my back to the window with the
three children in front of me as I spoke, and wondered if
they could see through the window behind me the large
municipal cemetery in the distance beyond the hospital car
park.

I concluded the meeting by telling them that there was no
need to make an immediate decision. I gave them my secretary's phone number and suggested they let me know what
they wanted to do in due course. They trooped out, and I
removed the three chairs and then went to collect the next
patient from the waiting area. I heard a few days later from
my secretary Gail that they had decided – how much persua-
sion the patient herself had required I do not know – that I
should operate.

She was admitted to the hospital for the operation three
weeks after the outpatient appointment. The evening before
the operation, however, the anaesthetist – a rather young
and inexperienced one – had requested a test called an echo-
cardiogram. She might, the anaesthetist said, have a problem with her heart because of her age, although she had no
symptoms of heart disease. This test was almost certainly
unnecessary but as a surgeon, with only minimal knowledge
of anaesthesia, I was in no position to argue. I told my juniors to beg the cardiac staff to get the test done first thing
the next morning. Instead of operating I had therefore spent
much of the day dozing angrily on the sofa in the surgeon's
sitting room, watching the dull sky through the high, view-
less windows, waiting for the test to be done. The occasional
pigeon flew past, and sometimes I could see airliners in the

distance nosing their way through the low clouds towards Heathrow.

The test, despite my juniors' pleas, was not done until four in the afternoon. Since the operation might well take several hours and I am only supposed to operate on emergencies out of hours I had to explain to the distressed and tearful patient, when she finally arrived outside the operating theatres in a wheelchair with an angry daughter, that I would have to cancel the operation. I promised to put her first on my next list so she was wheeled back to the ward and I bicycled home in a bad temper. Adding her to my next list would very probably mean having to cancel some of the other operations already planned for the day.

After I had discussed her case with the junior doctors at the Monday morning meeting I went to the reception desk outside the operating theatres. The anaesthetist – a different one from the one who had ordered the echo-cardiogram – was standing there with my registrar Mike, who looked gloomily at me.

'Mrs Seagrave's growing MRSA on the swabs which were taken when she was admitted last week and her operation was cancelled,' Mike said. 'The theatre will have to have a one-hour clean after her operation. We won't be able to get the list done if we do her first so I rearranged the list so that she's at the end of the list.'

'Well, I suppose we'll have to break my promise to her that we'd do her first,' I replied. 'It doesn't make much sense though does it? They test her for MRSA the day before the operation and don't get the results back for several days. If we had operated as planned last week we wouldn't have done a one-hour clean would we?'

'Mrs Seagrave's daughter was threatening to sue us last night,' he said. 'She said we were all hopelessly disorganized.'

'I'm afraid she's right but suing us isn't going to help, is it?'

'No,' he replied. 'It just puts one's back up. And it's quite upsetting.'

'Why the fuss?'

'The anaesthetist turned up and said she'd have to be cancelled.'

'Oh for Christ's sake, why?' I exploded.

'Because she's been put at the end of the list and therefore we won't finish before five o'clock.'

'Which bloody anaesthetist?'

'I don't know. A slim blonde – I think she's the new locum.'

I walked the few feet to the anaesthetic room and put my head through the door. The anaesthetist Rachel and her junior were leaning against the side of the worktop lining the wall of the anaesthetic room, drinking coffee from polystyrene cups, waiting for the first patient to arrive.

'What's all this about cancelling the last case?' I asked. The anaesthetist was indeed a new one – a locum recently appointed to replace my regular anaesthetist who was on maternity leave. We had done a few lists together and she had seemed competent and pleasant.

'I'm not starting a big meningioma at 4 p.m.,' she declared, turning towards me. 'I've got no childcare this evening.'

'But we can't cancel it,' I protested. 'She was cancelled once already!'

'Well I'm not doing it.'

'You'll have to ask your colleagues then,' I said.

'I don't think they will, it's not an emergency,' she replied in a slow and final tone of voice.

For a few moments I was struck dumb. I thought of how until a few years ago a problem like this would never have arisen. I always try to finish the list at a reasonable time but in the past everybody accepted that sometimes the list would have to run late. In the pre-modern NHS consultants never counted their hours – you just went on working until the work was done. I felt an almost overwhelming urge to play the part of a furious, raging surgeon and wanted to roar out, as I would have done in the past:

'Bugger childcare! You'll never work with me again!'

But it would have been an empty threat since I have little say in who anaesthetizes my patients. Besides, surgeons can no longer get away with such behaviour. I envy the way in which the generation who trained me could relieve the intense stress of their work by losing their temper, at times quite outrageously, without fear of being had up for bullying and harassment. I spun on my heel and walked down the corridor trying to work out how to solve the problem. The solution appeared immediately in the form of Julia the bed manager who was coming down the theatre corridor looking for me.

'We've admitted the two routine spines to the Day Room for your list today but we haven't got any beds to put them in afterwards, there were so many emergency admissions last night. What do you want to do?' she asked, looking strained. She was clutching her diary with its long list of patients who needed to be admitted, discharged or transferred and the phone numbers of the bed managers in other hospitals, who would probably be equally stressed and would be reluctant to accept them because they also were short of beds.

'If we haven't got any beds to put them in after the operation,' I said, inwardly rejoicing as this meant Mrs Seagrave's operation would now start early enough to mean I might finish by five o'clock, 'I can't very well operate, can I? You'll have to send them home. At least they're fairly minor ops.'

So the operating list was now sufficiently truncated. Two patients, starved from midnight in preparation for their frightening operations would be given a cup of tea in consolation and sent home.

I reluctantly walked to the Day Room on the ward where the patients to be operated upon that day were waiting. Since the hospital is chronically short of beds, more and more patients are brought in for surgery on the morning of their operation. This is standard practice in private hospitals and works very nicely since there will be a room and bed into which each patient can be put. In an over-stretched hospital like mine, however, this is not the case, so when I entered the small Day Room I found fifteen patients, all awaiting major surgery, all squeezed into a room the size of a small kitchen, still wearing their coats, wet with the February rain, which steamed in the cramped conditions.

Mike was on his knees in front of the patient who would be first on the list, since Mrs Seagrave's operation would now have to be done in the afternoon. He was explaining the consent form to him. He has rather a loud voice so all the other patients must have heard what he was saying.

'I have to warn you that there are some risks to the operation and these include death, a major stroke, major haemorrhage or serious infection. Just sign here, please.' He handed the consent form – a document that has become so complicated of late that it even has a table of contents on its front cover – to the patient with a pen and the man quickly scribbled his signature without looking at it.

I apologized to the two women whose spinal operations had been cancelled. I explained that there had been several emergency admissions during the night and they politely nodded their understanding, though I could see that one of them had been crying.

'We'll try to bring you back as soon as possible,' I said. 'But I'm afraid that at the moment I don't know when that will be.'

I dislike telling patients that their operation has been cancelled at the last moment just as much as I dislike telling people that they have cancer and are going to die. I resent having to say sorry for something that is not my fault and yet the poor patients cannot very well be sent away without somebody saying something.

I spoke briefly to the man with facial pain whose case I would do first and then to Mrs Seagrave, who was waiting in a corner with her daughter beside her.

'I'm so sorry about last week,' I said. 'And I'm sorry that I can't do your operation first but I promise to get it done this afternoon.' They looked a little dubiously at me.

'Well, let's hope so,' said her daughter with a grim expression. I turned to face all the patients crammed into the little room.

'I'm sorry about all this,' I said to them, waving my arm round the crowded room, 'but we're a bit short of beds at the moment.'

As I said this, suppressing the urge to deliver a diatribe about the government and hospital management, I wondered, once again, at the way in which patients in this country so rarely complain. Mike and I left for the theatres.

'Do you think I have said sorry enough?' I asked him.

'Yes,' he replied.

The first case was a microvascular decompression – known as an MVD for short. It was the same operation as the one I had been filmed carrying out in Kiev. The man had suffered from trigeminal neuralgia for many years and the standard pain-killing drugs had become increasingly ineffective. Trigeminal neuralgia is a rare condition – victims suffer excruciating spasms of pain in one side of the face. They say it is like a massive electric shock or having a red hot knife pushed into their face. In the past, before effective treatment became available, it was well-recognized that some people who suffered from it would commit suicide because of the pain. When I introduced the operation to Ukraine in the 1990s several of the patients I treated, since they could not afford the drug treatment, told me that they had indeed come close to killing themselves.

The operation involves exposing one side of the brain through a very small opening in the skull behind the ear and gently displacing a small artery off the sensory nerve – the trigeminal nerve – for the face. The pressure of the artery on the nerve is responsible for the pain though the exact mechanism is not understood. It is fairly exquisite microscopic surgery but, provided you know what you are doing, technically straightforward. Although Mike had been right to frighten the man with the consent form – and I had mentioned the same risks to him when I had seen him in my outpatient clinic a few weeks earlier – I have had only a few problems out of several hundred such operations and I did not seriously expect any difficulties.

Once I had entered his head and started to use the operating microscope I found an abnormally large vein blocking access to the trigeminal nerve. When I started to approach the nerve, deep in the part of the skull known as the cerebellopontine angle, the vein tore and a torrential haemorrhage of dark purple venous blood resulted. I was operating at a depth of six or seven centimetres, through a two-centimetre diameter opening, in a space only a few millimetres across, next to various vital nerves and arteries. The bleeding hides everything from view and you have to operate by blind reckoning, like a pilot lost in a cloud, until you have controlled the bleeding point.

'Suction up!' I shouted to the circulating nurse, as I tried to clear the blood with a microscopic sucker and identify from where the bleeding was coming.

It was not exactly a life-threatening emergency but it proved very difficult to stop the bleeding. You have to find the bleeding point and then pack it off with small pieces of haemostatic gauze which you press on with the tips of microscopic instruments which have angled handles so that your hands don't block the view, waiting for the vein to thrombose.

'It's not cool to lose your cool over venous haemorrhage,' I said to Mike as I gazed a little anxiously into the swirling pool of blood through the microscope. 'It will always stop with packing.' But as I said this I began to wonder if this might prove to be my second fatality with this operation. More than twenty years ago I had operated on an elderly man with recurrent trigeminal neuralgia and he had died from a stroke several weeks later as a result of the operation.

After twenty minutes, despite my efforts, the large sucker bottle at the end of the table was filled to its brim with dark red blood and Jenny, the circulating nurse, had to change it for an empty one. The patient had lost a quarter of all the blood circulating in his body. Eventually, with my instruments pressed against it, the packed off vein sealed and the bleeding stopped. As I stood there, my hands immobile as they held the microscopic instruments pressed onto the ruptured vein, I was certainly worried about the bleeding but I was equally worried that there now would not be enough time to get Mrs Seagrave's operation done. The thought of cancelling her operation a second time, and having to face her and her daughter again, was not a good one. Aware that I was starting to feel under pressure of time I felt forced to take even longer than perhaps was necessary to make sure that the bleeding had stopped. If it started again after I had closed his head, the result would almost certainly be fatal. By two o'clock I was happy with the haemostasis, as surgeons call the control of bleeding.

'Let's send for the next case,' I said to the anaesthetist, 'You've got an experienced registrar with you so she can

start the next case in the anaesthetic room while we're finishing in here.'

'I'm afraid we can't,' she replied, 'we have only one
ODA' – ODAs being the technicians who work with the
anaesthetists.

'Oh bloody hell, just send for the patient can't you?'

'The ODA manager has made a new rule that you cannot
start the next case until the first case is off the table. It's not
safe.'

I groaned and pointed out that we had never had any
problems overlapping cases in the past.

'Well, there's nothing you can do about it. Anyway, you
should plan more realistic operating lists.'

I could have explained that there was no way in which I
could have predicted the unusual bleeding. I could have explained that if I only planned operating lists that allowed for
the unexpected I would hardly get any work done at all. But
I said nothing. It was now unlikely that we would get Mrs
Seagrave's operation started in less than an hour after finishing the first case. I was going to have to operate in a hurry
if I was to finish by 5 p.m., something I hate doing. If the
operation did go on beyond 5 p.m. the theatre staff would
have to stay, of course, but if I 'over-run' too often it means
that in future it will become even more difficult to start cases
towards the end of the day. The thought of cancelling the
operation yet again, however, was even worse.

We finished the first case, and the anaesthetist started to
wake the man up.

'I think we can send now,' she said to one of the nurses,
who went outside to pass the message on. I knew that there
would be some delay before Mrs Seagrave was on the table
so I went down to my office to get some paperwork done. I
returned to the theatres after twenty minutes and looked into
the anaesthetic room expecting to see the anaesthetists busy

at work on Mrs Seagrave. To my dismay I saw that the room was empty apart from an ODA, whom I did not recognize.

I asked him what had happened to the patient but he just shrugged and said nothing in reply so I headed off to the Day Room to see what had happened to Mrs Seagrave.

'Where's Mrs Seagrave?' I asked the nurse.

'She's gone to get changed.'

'But why wasn't she changed already?'

'We're not allowed to.'

'What do you mean?' I asked in exasperation. 'Who doesn't allow it?'

'It's the government,' the nurse replied.

'The government?'

'Well the government says we can't have patients of different sexes sitting in the same room in theatre gowns.'

'Why not put them in dressing gowns?'

'We suggested that ages ago. The management said the government wouldn't allow it.'

'So what should I do? Complain to the prime minister?' The nurse smiled.

'Here she is,' she said, as Mrs Seagrave appeared, being pushed along the corridor in a wheelchair by her daughter. She was dressed in one of those undignified hospital gowns that scarcely cover one's buttocks, so perhaps the government was right after all.

'She had to change in the toilet,' said her daughter, rolling her eyes.

'I know. There are no separate facilities for the patients who come in on the morning of the operation,' I said. 'Anyway, we're running out of time. I'll take her to theatre myself.' So I took hold of the wheelchair and rolled her rapidly down the corridor.

The ward nurse came running down the corridor after me clutching Mrs Seagrave's notes.

By now it was three o'clock and the anaesthetist was look-ing distinctly unhappy.

'I'll do it all myself,' I hurriedly assured her. 'Skin to skin.' Mike was disappointed that I would be elbowing him aside – earlier in the day I had told him that I would assist him doing it. Now he would have to assist me.

'It looks very straightforward. It's going to be easy,' I added. This was a lie and I did not expect Rachel to believe it. Few anaesthetists believe what surgeons tell them.

And so, at half past three, we started.

Mike bolted the patient's head to the operating table and shaved the left side of her head.

'These are operations where one really doesn't know what's going to happen,' I muttered to Mike, not wanting Rachel to hear. 'She might bleed like a stuck pig. The tumour might be horribly stuck to the brain so it will take hours and at the end we're left with the brain looking a horrible mess and she's crippled, or the tumour might just jump out and scamper round the theatre.'

With scalpels, drills and clips, together we worked our way steadily through the scalp and skull of the late eminent gynaecologist's widow. After forty minutes or so we were opening the meninges with a small pair of scissors to expose her brain and the meningeal tumour pressing into it.

'Looks pretty promising,' said Mike, bravely hiding his disappointment at not doing the operation himself.

'Yes,' I agreed. 'Not bleeding much and looks as though it will suck nicely.' I picked up my metal sucker and stuck it into the tumour. It made an unattractive sucking sound as the tumour started to disappear, peeling gently off the brain as it shrank.

'Awesome!' said Mike. After a few minutes I shouted out happily to Rachel: 'Forty minutes to open her head. Ten minutes to remove the tumour! And it's all out and the brain looks perfect!'

'Wonderful,' she said, though I doubted if I was forgiven.

I left Mike to close up the old lady's head and sat down in a corner of the theatre to write an operating note. It took another forty minutes to finish the operation and the patient was being wheeled off to the Intensive Care Unit by 5 p.m.

Mike and I left the theatres and walked round the wards to see our inpatients. Apart from the two surgical cases we had just done there were only a few patients, recovering un- eventfully from relatively minor spinal operations done two days earlier so the ward round took only a few minutes and we ended up on the ITU. Examining patients at the end of the operating list, making sure that they are, as the jargon has it, 'awake and fully orientated with a GCS of 15', is an important part of the neurosurgeon's day.

Mrs Seagrave was sitting half-upright in her bed, with drip-stands and syringe pumps and monitors with flashing displays beside her. With so much technology it is hard to believe that anything can go wrong but what really matters is that a nurse wakes the patient up every fifteen minutes to make sure they are alert and not slipping into a coma caused by post-operative bleeding. A nurse was cleaning blood and bone dust from her hair. I had finished the operation in a hurry and had forgotten to wash and blow-dry her hair, something I usually do with female patients.

'It all went perfectly,' I said, leaning slightly over her from the side of the bed. Mrs Seagrave reached out for my hand and held it tightly.

'Thank you,' she said, in a voice a little hoarse from the anaesthetic tube.

'All out, and definitely benign,' I said. I turned away and went to see the man with the trigeminal neuralgia who was in the bed next to hers. He was asleep and I shook him gently. He opened his eyes and looked a little groggily at me.

'How's your face feel?' I asked.

He cautiously touched his cheek. Before the operation doing this would have caused him the most terrible agony.

He looked surprised and prodded his cheek a little more forcefully.

'It's gone,' he said, in an awe-struck voice and smiled happily. 'That's wonderful.'

'The op went fine,' I said. 'Definite artery on the nerve. Consider yourself cured.' I saw no need to mention the awful bleeding.

I went down the stairs to my office to see if there was any more paperwork to be done but just for once Gail had left my office empty. It had been a good day. I had not lost my temper. I had finished the list. The patients were well. The pathology had been benign. I had been able to cancel the two spines at the beginning of the list rather than at the end. There were no major problems with the patients on the wards. What more could a surgeon want?

On my way out I passed Anthony, who was coming in for the evening shift. I asked about the old lady with a chronic subdural who wanted to die.

'I think they operated,' he said. He headed off to the wards and I walked out into the night. Mrs Seagrave's daughter was standing outside the hospital entrance, beside the railings where I padlock my bicycle, smoking a cigarette.

'How did it go?' she asked when she saw me.

'It went perfectly,' I replied. 'She might be a bit confused for a few days but I think she'll make a very good recovery.'

'Well done!' she said.

I told her that it was largely a matter of luck but she

probably didn't believe me – they never do when an operation has gone well.

'I'm sorry I lost my temper with your registrar yesterday...' she began.

'Don't think about it,' I replied cheerfully. 'I was an angry relative myself once.'

8

CHOROID PLEXUS PAPILLOMA

n. a benign tumour of the choroid plexus, a structure made from tufts of villi within the ventricular system that produces cerebrospinal fluid.

Thirty years ago British hospitals always had a junior doctor's bar where you could go for a drink at the end of a long day, or where – if you had any free time – you could spend the evening smoking and drinking when on call, or playing on the Space Invaders or Pacman machines in a corner of the room.

I was working as a gynaecological houseman, and had only been qualified as a doctor for four months. It was to be eighteen months before I saw the operation that convinced me to become a neurosurgeon. I was standing at the bar one evening, drinking beer and gossiping with colleagues, probably discussing patients and their illnesses in that slightly swaggering way that young doctors have when talking to each other. I was probably also feeling a little guilty about not returning home more promptly to see my wife Hilary and our three-month-old son William when my bleep announced an outside call. I found the nearest phone to be told by Hilary, who sounded desperate, that our son had been admitted to the local hospital, seriously ill, with some kind of problem in his brain.

I remember very clearly how I ran from the hospital to the underground station, and once off the train, sick with anxiety, sprinted through the dark and deserted back streets of Balham – it was winter and it was already late in the evening – to the local hospital. There I found a distraught Hilary in a quiet side-room, our baby son sleeping restlessly in her arms, and a consultant paediatrician who had waited for my arrival. He told me that William had acute hydrocephalus and would be transferred to the children's hospital at Great Ormond Street the next day for a brain scan.

My wife and I spent the next few weeks in that strange world one enters when you fear for your child's life – the outside world, the real world, becomes a ghost world, and the people in it remote and indistinct. The only reality is intense fear, a fear driven by helpless, overwhelming love.

He was transferred on a Friday afternoon – never a good time to fall seriously ill – and a brain scan was organized. Because I was a doctor myself, and the junior doctor looking after William turned out, by an odd coincidence, to be an old school-friend of Hilary's, I was allowed to stand in the control room for the scanner. It was strange to hear the two radiographers happily chatting about a party they had been to, detached and uninterested in the little baby wrapped in a blanket who could be seen through the control room window lying in the big mechanical doughnut of the machine with his mother, looking drawn and desperate, sitting beside him. I watched the images appearing on the computer screen as the scanner slowly worked its way up through William's head. They showed acute hydrocephalus and a tumour immediately in the centre of his brain.

He was taken back to the ward from the scanner. I was told the consultant surgeon would be coming to see him later. William was now obviously – or at least it looked obvious to me – unconscious and very ill but I was assured by the surgical registrar that he was just sleeping off the sedation he had been given for the scan. The afternoon passed and it became dark outside. We were then told that possibly the consultant surgeon would not be coming until the following Monday. In a kind of fugue state, I wandered around the hospital's long corridors, now largely empty, helplessly trying to find the consultant – a man who seemed to have become as mythical as the neurosurgeons in my own hospital – and eventually, despairing, unable to stand it any longer, I abandoned my wife and child and went home, where I smashed a kitchen chair in front of my alarmed parents and swore to sue the hospital if William came to any harm.

While I was failing so dismally to cope with the situation the surgeon had, I later learned, appeared, taken one look at William and ushered Hilary out of the room. He inserted emergency drains into William's brain through the fontanelle to relieve the build-up of pressure – at least in retrospect I could claim to have been right to have been so frightened. We were told that an operation to remove the tumour would take place five days later. The five days were torture.

When I drove home the night before the operation, a black cat suddenly ran out in front of my car, a few hundred yards away from our home. The wheels of my car ran straight over it. I had never killed any animal in this way before and have never done so since. I got out of the car and went to look at the poor creature. It lay in the gutter, obviously dead, its mouth and eyes open, bared at the moon above in the clear winter sky. I remembered how the name tag around William's little wrist had a cat's face on it, since he was in a children's hospital, and they like to do things that way. I am not a superstitious man but I found this very frightening.

William underwent surgery on a Wednesday morning. Hilary and I spent many hours pacing around central London while the operation went on. It was a useful lesson for me, when I became a fully trained surgeon myself, to know how much my patients' families suffer when I am operating.

The operation was a success and William survived, since the tumour proved to be a benign choroid plexus papilloma even though the pathology report had reported it to be malignant. I came to realize later that few brain tumours at that age are benign, and that even with the benign tumours the risks of surgery in such young children are immense. Years afterwards, when training as a paediatric brain surgeon myself, I watched a child bleed to death in the very same operating theatre where my son had been treated, as my boss – the very surgeon who had saved my son's life – now failed with a similar tumour.

Anxious and angry relatives are a burden all doctors must bear, but having been one myself was an important part of my medical education. Doctors, I tell my trainees with a laugh, can't suffer enough.

LEUCOTOMY

n. the surgical cutting of tracts of white nerve fibres in the brain; orig. spec., prefrontal lobotomy; an instance of this.

My department is unusually fortunate in having a sitting room for the surgeons beside the operating theatres. The room is furnished with the two large red leather sofas that I bought shortly after we moved from the old hospital. When our department was moved from the old hospital to a newly built block at the main hospital some miles away, the entire second floor of the new building was dedicated to neurosurgery. As time passed, however, the management started to reduce our facilities and one of the neurosurgical theatres became a theatre for bariatric surgery – surgery for morbidly obese people. The corridors and rooms were starting to fill with unfamiliar faces and patients the size of small whales being wheeled past on trolleys. The department no longer felt like our home and I feared that I was starting to develop the slightly alienated, institutionalized outlook that afflicts so many of the staff working in huge, modern hospitals.

I was sitting in the red leather sofa room one day reading a book while my registrar started a case. We had taken to keeping the door of the room locked, as there were so many strangers now present in the theatre department. As I sat on my sofa somebody started knocking and shaking the door.

I felt increasingly foolish sitting there refusing to open the door. Eventually, to my dismay, the door was forced open and four doctors – none of whom I recognized – burst into the room with sandwiches in their hands. Embarrassed, I stood up.

'This is the neurosurgical office!' I said, feeling like a pompous fool. 'You're not welcome here!'

They looked at me in surprise.

'The management said all the facilities would be shared,' one of them said, looking at me in disgust.

'Well the management never discussed it with us,' I replied. 'If you had your own office, wouldn't you resent it if other people barged in without asking?'

'We're surgeons,' one of them said, shrugging, but they left the room, and I also left, too upset to remain, yet determined to maintain what little remained of our neurosurgical territory.

I joined my registrar in the operating theatre, where I took over the operation. It was an unusually difficult case and I damaged the nerve for the left side of the patient's face as I removed the tumour. Perhaps this was going to happen anyway – it is called a 'recognized complication' of that particular operation – but I know that I was not in the right state of mind to carry out such dangerous and delicate surgery, and when I saw the patient on the ward round in the days afterwards, and saw his paralysed face, paralysed and disfigured, I felt a deep sense of shame. It is little consolation that my colleagues and I have been left undisturbed in the red leather sofa room, our little oasis, ever since, although I believe I have become an object of deep dislike among many of the other surgeons in the hospital.

For reasons that have never been determined, all the windows in the offices for the operating theatre block, including the room with the red leather sofas, are five feet off the floor.

All you can see through them, once you sit down, is the sky, with an occasional plane on its way to Heathrow or more often a pigeon, or sometimes a sea gull and very occasionally a kestrel. I have spent many hours lying on the longer of the two sofas, reading medical journals, struggling to keep awake, waiting for the next case to begin, watching the dull clouds through the high windows. In recent years the delays between finishing one operation and starting the next have become longer and longer. The trouble is that we cannot start the operation until we know that there is going to be a bed into which to put the patient after the operation, and this is often not the case. The stream of initiatives and plans and admonitions from the government and management that we must work ever more efficiently feels like a game of musical chairs – the music is constantly being changed, indeed with the latest round of reforms the government has even changed the orchestra – but there are always more patients than beds and so I spend many hours lying on a sofa, staring gloomily at the clouds, watching the pigeons hurrying by.

I was lying on the sofa waiting for the next case to start and dozing over a book. My colleague who operates on the same days as I do was sitting in a chair, waiting like myself for his next case to be anaesthetized.

'I see that we have been told that the whole culture of the NHS must change – after all those patients died in Stafford. What a whitewash. It's all about who's in charge,' he said.

I remembered when, still a student, I had spent several months working as a nursing assistant on the long-term psycho-geriatric ward of one of the huge long-term psychiatric hospitals that used to surround London. Most of the patients were profoundly demented. Some had come from the outside world with degenerative brain diseases, some were schizophrenics who had already spent most of their lives in the hospital and were now sinking to the end of their lives. To go to work at seven in the morning to be faced by a room of twenty-six doubly incontinent old men in beds is an education of sorts, as it was to wash them and shave them and feed them, and pot them, and strap them into geriatric chairs. I met some nurses who were utterly unsuited to the work and others who were quite remarkably patient and kind, in particular a wonderful West Indian man called Vince Hurley, who was the charge nurse for the ward. It was miserable work, with little reward, and I learned much about the limitations of human kindness, and in particular my own.

I was told that in the nineteenth century when the severe and prison-like hospital had been built there had been a hospital farm on the extensive grounds and that the patients had worked on the farm, but when I was there the grounds were just a series of wide, empty fields. Rather than carry out farm work outside, some of the patients now received something called Occupational Therapy. This involved three occupational therapists – stout middle-aged ladies in maroon-coloured housecoats – leading out a straggling line of demented old men onto the fields around the hospital twice a week. It was 1976, the year of the great drought, and the hospital grounds had been burnt brown and yellow, and the patients' faces had been burnt red since most of them were on the anti-psychotic drug Largactil, which is a photo-sensitizer. The patients were given a football and left to their own devices – most of them sat down and stared into space. The three therapists also sat down. One particularly catatonic patient – he had been lobectomized many years ago – could sit immobile for hours on end and served as a backrest for one of the therapists, as she sat on the burnt grass, her back resting comfortably against his as she did her knitting. He was called Sydney, and he was famous for having enormous genitals. I had been summoned by the other nurses, at washing time, on my very first day at work, to admire Sydney's equipment, as he lay catatonically in the bath.

It was while working here that I first came across the name of the famous neurosurgical hospital where I was to train and eventually become the senior consultant neurosurgeon myself. In the 1950s many of the patients I was now looking after – like the catatonic Sydney – had been sent to that hospital and subjected to the psychosurgical procedure known as a frontal lobectomy or leucotomy. It was a fashionable treatment at the time for schizophrenia and was supposed to turn agitated, hallucinating schizophrenics into calmer, happier people. The operation involved severing the frontal lobes from the rest of the brain with a specially shaped knife and was completely irreversible. Fortunately it was rendered obsolete by the development of phenothiazine drugs such as Largactil.

The lobectomized men were, it seemed to me, some of the worst affected of the patients – dull and apathetic and zombie-like. I was shocked to find, when surreptitiously looking at their notes, that there was no evidence of any kind of follow-up or post-operative assessment. In all the patients who had been lobectomized there would be a brief note stating 'Suitable for lobectomy. For transfer to AMH'. The next entry would read 'Returned from AMH. For removal of black silk sutures in nine days', and that was it. There might be the occasional entry years later stating, for instance 'Called to see. Fight with other patient. Scalp laceration sutured', but other than the notes made at the time of the patient's first admission to the hospital, usually with an episode of acute psychosis, the medical notes were empty even though the patients had been in the hospital for many decades.

Two years earlier a Royal Commission on Psychiatric Care

had been established in response to an outcry in the press over accusations of brutality made by a student who, rather like myself, had worked as a nursing assistant in a long-term psychiatric hospital. It was why I was viewed with considerable suspicion by the other hospital staff when I arrived and it took me some time to persuade them that I was not spying on them. I suspect that some things were kept hidden from me, but I saw little, if any, actual cruelty when I was there.

I was surprised one morning, when spooning gruel into an old man's edentulous mouth, to see the nursing officer come into the dining room. He told me that I had the afternoon off, though gave me no reason. He had brought with him a large laundry bag full of worn but clean old suits, some of them pinstriped, and much underwear. The patients were all doubly incontinent so we kept them all in pyjamas as it was easier to change them and keep them clean, but my fellow nurses and I were told that all the patients were now to be dressed in suits and underwear. So our poor, demented patients were all dressed up in sagging, second-hand suits, and put back in their geriatric chairs and I went home. When I clocked in for the late shift next day I found the patients all back in pyjamas and the ward back to normal.

'The Royal Commission came yesterday,' Vince said to me with a grin 'They were very impressed by the suits. The nursing officer didn't want you around in case you said the wrong thing.'

Vince was one of the most impressive people I have met in my long medical career. To work on that ward, with those hopeless cases, to treat them with such kindness and tact, was remarkable. Sometimes he would stand behind one of the babbling, demented incontinent old men, and lean his hands, sleeves of his white coat rolled up, on the high back of the patient's chair.

'What's it all about?' he would say with a sigh. 'That's

what I want to know. What's it all about?' and we would laugh, and get on with the day's work – feeding the patients, washing the patients, lifting them on and off the toilet, and eventually putting them to bed for the night.

Thirty-five years later the hospital is still there but the grounds have been sold and have become a smart golf course. The patients I looked after must all have died a long time ago.

'What are you reading?' my colleague asked, seeing that there was a book on my lap.

'Something incomprehensible about the brain,' I said, 'written by an American psychologist who specializes in treating obsessional compulsive disorder with group therapy based on combining Buddhist meditation with quantum mechanics.'

He snorted. 'How fucking ridiculous! Didn't you once do psychosurgery for OCD?'

It was true. I had inherited the operation from my predecessor but had been happy to abandon it. It involved making lesions in the caudate nucleus and cingulate gyrus of the frontal lobes – a sort of micro-lobectomy without its awful effects. The psychiatrists told me that the operation really did work. It had all seemed rather like guesswork to me but recent high-tech functional scanning in OCD shows that these are indeed the areas involved. Psychosurgery was banned by law in California so a few desperate Californians who had become quite suicidal because they couldn't stop washing their hands – fear of dirt being one of the commonest problems with OCD – used to come to this country for treatment. I remembered how one of them had to put on three pairs of gloves before he could touch the pen I handed him for signing the consent form that allowed me to burn a few holes in his brain. As I told my colleague about my experience of psychosurgery a nurse came into the room.

'Mr Marsh,' she said, looking disapprovingly at me as I lay sprawled on the sofa in my theatre pyjamas, 'the next patient says his tumour is on the right side but the consent form says for surgery on the left.'

'Oh for God's sake,' I said. 'He's got a left parietal tumour and has got left-right confusion as a result. You might like to know it's called Gerstmann's syndrome. He's the last person to ask where to operate! He's been thoroughly consented. I spoke to him myself last night. And the family as well. Just get on with it.'

'Some people don't think Gerstmann's syndrome really exists,' my colleague – who is very knowledgeable about such things – said from across the room.

'You must go talk to him,' the nurse said.

'This is ridiculous,' I grumbled as I rolled off the sofa. I walked the short distance to the anaesthetic room, through the theatre where Kobe the theatre porter was cleaning up after the first operation, mopping up blood, smeared in ragged lines on the floor. There was the usual pile of rubbish – several thousand pounds worth of single-use equipment scattered around the operating table, waiting to be bagged up and sent off for disposal. I pushed through the swing doors to the anaesthetic room where the old man was lying on a trolley.

'Mr Smith. Good morning!' I said. 'I gather you want me to operate on the right side of your head.'

'Oh Mr Marsh! Thank you for coming! Well I thought it was on the right,' he replied, his voice trailing off in uncertainty.

'Your weakness is on the right,' I said. 'But that means the tumour is on the left of your brain. Everything is crossed over, you know.'

'Oh,' he replied.

'Well, I'll operate on the right side if you want but would

you perhaps prefer me to decide on the side?'

'No! No!' he said, laughing 'You decide.'

'Well, left side it is then,' I said.

I left the anaesthetic room. The nurse would now tell the anaesthetist that she was allowed to start the operation. I went back to the red leather sofa.

Forty minutes later the nurse returned to say that the next patient was now anaesthetized and I sent my junior off to start the case. The junior doctors work such short hours that they are desperate for even the most basic surgical experience and I feel obliged to leave all of the opening and closing to them as this is a simple and relatively safe part of brain surgery, even though I would much prefer to do it myself. The intense anxiety I experience when supervising my juniors, however, so much greater than when I operate myself, means that I find it quite impossible to leave the theatres when they are operating on anything but the simplest cases, and there is far too much paperwork to bring up from my office, so I feel compelled instead to stay in the room with the red leather sofa.

I will wander in and out of the theatre and watch, a little jealously, what they are doing and only scrub up when the patient's brain is reached and the operation becomes more intricate and dangerous. The point at which I take over will depend on the experience of the trainee and the difficulty of the case.

'How's it going?' I will ask, as I walk into the theatre, putting on my reading glasses and a face mask, to peer into the wound.

'Fine, Mr Marsh,' my trainee will reply, wanting me to go away, and well aware of the fact that I would love to elbow him or her aside and take over the operation.

'You're sure you don't need me?' I will ask hopefully and will usually be assured that everything is under control. If this indeed appears to be the case I will turn away from the operating table with a sigh and walk the few yards back to the sitting room.

I stretched out on the sofa and carried on reading my book.

As a practical brain surgeon I have always found the philosophy of the so-called 'Mind-Brain Problem' confusing and ultimately a waste of time. It has never seemed a problem to me, only a source of awe, amazement and profound surprise that my consciousness, my very sense of self, the self which feels as free as air, which was trying to read the book but instead was watching the clouds through the high windows, the self which is now writing these words, is in fact the electrochemical chatter of one hundred billion nerve cells. The author of the book appeared equally amazed by the 'Mind-Brain Problem', but as I started to read his list of theories – functionalism, epiphenomenalism, emergent materialism, dualistic interactionism or was it interactionistic dualism? – I quickly drifted off to sleep, waiting for the nurse to come and wake me, telling me it was time to return to the theatre and start operating on the old man's brain.

10

TRAUMA •

n. any physical wound or injury.
psychol. an emotionally painful and harmful event.

I was early and had to wait for the junior doctors to arrive. The days of white coats are long gone and instead the juniors turn up in Lycra bicycling gear or, if they have been on duty overnight, in the surgical scrubs made popular by TV medical dramas.

'There was only one admission last night,' the on-call registrar said, sitting at the front of the room beside the computer keyboard. She was quite unlike the other trainees, who are usually full of youthful enthusiasm. She talked in an irritated and disapproving tone of voice. This invariably had a dampening effect on the meetings when it was her turn to present the cases. I had never understood why she wanted to train as a neurosurgeon.

'He's a forty-year-old man,' she said. 'Seems he came off his bike last night. He was found by the police.'

'Push-bike?' I asked.

'Yes. And like you he wasn't wearing a crash helmet,' she said, with a look of disapproval. As she talked she typed on the keyboard and the slices of a huge black-and-white brain scan started to appear, like a death sentence, out of the dark onto the white wall in front of us.

'You won't believe this,' one of the other registrars broke in. 'I was on yesterday evening and took the call. They sent the scan on a CD but because of that crap from the government about confidentiality they sent two taxis. Two taxis! One for the fucking CD and one for the little piece of paper with the fucking encryption password! For an emergency! How stupid can you get?'

We all laughed, apart from the registrar presenting the case who waited for us to calm down.

'The police said he was talking when they found him,' she went on, 'but when he was admitted to the local hospital he started fitting so he was tubed and ventilated and then scanned.'

'He's stuffed,' somebody called out from the back of the room as we looked at the scan.

'I hope he doesn't survive,' the on-call registrar suddenly said. I was very surprised since I knew from past experience that she believed in treating patients even with a hopeless prognosis.

I looked at the junior doctors in the front row.

'Well,' I said to one of them, a dark-haired girl who had only just started in the department, and who would only be with us for two months. 'There are many abnormalities on the scan. See how many you can identify.'

'There's a frontal skull fracture, and it's depressed – the bone's been pushed into the brain.'

'What's happened to the brain?'

'There is blood in it – contusions.'

'Yes. The contusions on the left are so big that it's called a burst frontal lobe. All that area of brain has been destroyed. And what about the other side?'

'There are contusions there as well, but not as big.'

'I know he was talking at first and in theory might make a good recovery but sometimes you get delayed intraparenchymal bleeding like this and the scan now shows catastrophic brain damage.'

'What's his prognosis?' I asked the registrar.

'Not good,' she said.

'But how much not good?' I asked. 'Fifty per cent? Ninety per cent?'

'He might recover.'

'Oh come off it! With both his frontal lobes smashed up like that? He hasn't got a hope in hell. If we operate to deal with the bleeding he might just survive but he'll be left hopelessly disabled, without language and probably with horrible personality change as well. If we don't operate he'll die quickly and peacefully.'

'Well, the family will want something done. It's their choice,' she replied.

I told her that what the family wanted would be entirely determined by what she said to them. If she said 'we can operate and remove the damaged brain and he may just survive' they were bound to say that we should operate. If, instead, she said 'If we operate there is no realistic chance of his getting back to an independent life. He will be left profoundly disabled. Would *he* want to survive like that?' the family would probably give an entirely different answer. What she was really asking them with the first question was 'Do you love him enough to look after him when he is disabled?' and by saying this she was not giving them any choice. In cases like this we often end up operating because it's easier than being honest and it means that we can avoid a painful conversation. You might think the operation has been a success because the patient leaves the hospital alive but if you saw them years later – as I often do – you would realize that the result of the operation was a human disaster.

The room was silent for a while.

'The decision has been made to operate,' the registrar

said stiffly. Apparently the patient was under the care of one of my colleagues and one of the unwritten rules of English medicine is that one never openly criticizes or overrules a colleague of equal seniority, so I remained silent. Most neurosurgeons become increasingly conservative as they get older – meaning that they advise surgery in fewer patients than when they were younger. I certainly have – but not just because I am more experienced than in the past and more realistic about the limitations of surgery. It is also because I have become more willing to accept that it can be better to let somebody die rather than to operate when there is only a very small chance of the person returning to an independent life. I have not become better at predicting the future but I have become less anxious about how I might be judged by others. The problem, of course, is that so often I do not know just how small the chance of a good recovery might be because the future is always uncertain. It is much easier just to operate on every case and turn one's face away from the fact that such unquestioning treatment will result in many people surviving with terrible brain damage.

We all filed out of the room and scattered over the hospital for the day's work – to the theatres, to the wards, to the outpatient clinic, to the offices. I walked down the X-ray corridor with my neuroradiological colleague. Neuroradiologists spend their day analysing brain and spinal scans but do not usually deal directly with patients. I think he had started his career in neurosurgery but was too gentle a soul to be a neurosurgeon and so had become a neuroradiologist.

'My wife's a psychiatrist, you know,' he said. 'When she was training she worked in a brain damage unit for a while. I'm with you on this one – so many of the head injuries have terrible lives. If neurosurgeons followed up the severe head injuries they treated I'm sure they'd be more discriminating in whom they operated upon.'

I went down to my office where I found my secretary Gail cursing her computer again while she tried to get onto one of the hospital databases.

I noticed a sheet of paper beside her keyboard printed in crude colours with flowery capitals.

'This certificate is presented ...' it began. It went on to state that Gail had attended something called a MAST Catchup Seminar.

'What this?' I asked pointing to the piece of paper.

'Mandatory and Statutory Training.' she said. 'It was a complete and utter waste of time. It was only bearable because some of your colleagues were there and spent the whole time taking the piss out of the lecturer who was completely useless. I was told afterwards his background was in catering – he didn't know anything he was talking about. He'd just been trained to say it. You're going today, have you forgotten?' she added in a mockingly disapproving voice. 'It's mandatory for all members of staff and that includes consultant surgeons.'

'Oh really?' I replied, but it was true that I had received a letter from the chief executive a few weeks earlier. The letter stated that it had been brought to his attention that I had not attended Mandatory and Statutory Training and that it was indeed mandatory and statutory that I attend. The fact that he had found the time to write to me showed that the MAST course was clearly of vital importance.

So I walked out of the hospital into the late August sunshine and made my way across one of the hospital's many car parks, narrowly missed being run over by a long line of wheelie bins being tugged round the perimeter road by a bored-looking man on a small tractor, and presented myself to the Training and Development Centre, a large and flimsy mobile home, the floor of which shook as I strode angrily down the corridor to the room where the MAST seminar was to be held. I was late and there were already about forty people sitting glumly at desks – a mixed group of nurses, cleaners, clerks and doctors and doubtless other members of the huge bureaucracy that forms an NHS Trust. I took a chair and sat in the far corner at the back. The lecturer – a young man with a trimmed ginger beard and shaven head – came and presented me with a folder entitled 'MAST Workbook'. I felt as though I was back in school and I refused to take it from him – so, with a sigh, he patiently left it on the floor beside me and returned to the front of the room and turned to face his audience.

The seminar was scheduled to last three hours and I settled down to get some sleep. The long hours I worked as a junior doctor in the distant past have taught me the art of getting to sleep virtually anywhere on any surface.

Halfway through there was a coffee break before we were to learn about Fire Drill and the Principles of Customer Care. As we walked out of the room I picked up a message on my mobile phone, which I had dutifully turned off. One of my patients on the women's ward was dying and the ward sister had rung to say that the family wanted to talk to me. So I returned to the hospital and went up to the ward.

The patient in question was a woman in her forties with breast cancer who had developed a secondary tumour in her brain. This had been operated on and removed by one of the senior trainees a week earlier but two days after the operation – which had gone uneventfully – she had suffered a major stroke. She was now dying. I had realized earlier in the week, with something of a shock, that nobody had spoken to the family about this. The surgeon who had done the operation was away on leave, as was my own registrar. I had been busy operating and none of the small army of shift-working junior doctors had felt sufficiently involved with the patient – whom they did not know – to talk to the family. I had therefore arranged to meet them at nine in the morning, forgetting that I was supposed to be attending the MAST seminar.

I found the patient's husband and elderly mother sitting sadly beside the woman's bed, squeezed into the narrow space between her bed and the next patient's in a six-bed bay. She was unconscious, and breathing heavily and irregularly. There were five other patients in the room, with only two feet or so between their beds, who could watch her slowly dying.

I hate breaking bad news to patients and their families in rooms like this, overheard by others, hidden by a flimsy curtain. I also hate talking to patients and their families – 'customers' as the Trust would have it – while standing, but there were no empty chairs in the bay so I stood unhappily above the dying woman and her family as I spoke to them. It seemed inappropriate to sit on the bed, besides I believe it is now banned by Infection Control.

'I'm very sorry I didn't talk to you earlier,' I began. 'I'm afraid she suffered a stroke after the operation. The tumour was stuck to one of the major arteries for her brain and when this happens, even though we are able to remove the tumour, sometimes a stroke occurs in the artery a few days later.'

The husband and mother looked at me silently.

'What will happen?' her elderly mother asked.

'Well,' I said hesitating, 'I think she will probably ...' I hesitated again, and dropped my voice, acutely aware of the other patients listening and wondering whether to use one of the many euphemisms for death. 'I think she will probably die, but I just don't know if it will be within the next few days or take longer.'

Her mother started crying.

'It's every parent's worst nightmare to outlive their children,' I said.

'She was my only child,' her mother replied through her tears. I reached out to put my hand on her shoulder.

'I'm so sorry,' I said.

'It's not your fault,' she replied. There was nothing else to be said so after a while I went to find the ward sister.

'I think Mrs T is dying,' I said. 'Can't we put her in a side room?'

'I know,' the sister said, 'We're working on it but we're desperate for beds at the moment and we'll have to shift an awful lot of patients around.'

'I'm at the MAST meeting all about Customer Care this morning,' I said.

The ward sister snorted. 'We just give such crap care now,' she said with feeling. 'It used to be so much better.'

'But the patients are always telling me how good it is here,' I said, 'compared to the local hospitals.' She said nothing and hurried away, forever busy.

I returned to the Training and Development Centre. The second session had already started. The PowerPoint presentation now showed a slide with a long list of the Principles of Customer Service and Care.

'Communicate effectively,' I read. 'Pay attention to detail. Act promptly.' We were also advised to develop Empathy.

'You must stay composed and calm,' Chris the lecturer told us. 'Think clearly and stay focused. Your emotions can affect your behaviour.'

How strange it is, I thought as I listened to him talking, that after thirty years of struggling with death, disaster and countless crises and catastrophes, having watched patients bleed to death in my hands, having had furious arguments with colleagues, terrible meetings with relatives, moments of utter despair and of profound exhilaration – in short, a typical neurosurgical career – how strange it is that I should now be listening to a young man with a background in catering telling me that I should develop empathy, keep focused and stay calm. As soon as the signing-out register had been passed around, and I had signed it, thus confirming that the Trust could now state that I had been trained in Empathy and Self-control, and the classification of Abuse and of Fire Extinguishers, in addition to many other things I had already forgotten, I charged out of the room despite Chris' protests that he had not yet finished.

Next morning, while I was telling Gail about my training, a junior doctor came to the door. He looked anxious and unhappy. He was one of the doctors on the neurology ward – a ward for people with illnesses of the brain who do not need surgical treatment. These are illnesses like multiple sclerosis or Parkinson's disease, or strange and obscure diseases, sometimes untreatable, which neurologists find deeply fascinating, and which they collect like rare butterflies and report in their journals.

'I'm sorry to interrupt you ...' he began.

'Not at all,' I replied, pointing to the piles of notes and paperwork on my desk and on the floor around me, 'I'm only too happy to be distracted.'

'We admitted a fifty-nine-year-old woman over the weekend with progressive dysphasia and then fits and on the scan it looks as though she's got ADEM.'

'ADEM? It doesn't sound very surgical,' I said.

'Acute disseminated encephalo-myelitis,' he replied – in other words, a sudden and catastrophic inflammation of all of the brain and spinal cord.

I told him that I didn't think surgery would be much help.

'Yes, but she's gone off this morning and blown her left pupil and the scan shows diffuse swelling. We thought she might need decompressing.'

I reached for my computer keyboard. It sounded as though her brain had become so swollen that she was starting to die from the build-up of pressure in her head, the swollen brain being trapped, so to speak, within the skull. A 'blown' pupil – meaning that the pupil of one of her eyes had become very large and was no longer contracting if a light was shone in it – is the first sign of what can be a rapidly fatal process. The fact that she had 'gone off' – that she had become unconscious – meant that if something was not done very quickly to reduce the pressure in her head she would die within the next few hours, or even less.

The scan showed that all of her brain, but especially the left side, was dark with severe swelling, the medical word for this being cerebral oedema. The oedema was a reaction to the ADEM, although the actual cause of ADEM is not known.

Some parts of the brain can be removed without leaving the patient disabled, but if I removed the swollen part of this woman's brain she would be left hopelessly disabled, unable to talk or even understand language.

'What about a decompressive craniectomy?' the neurology junior asked. This is an operation where you remove the top of the patient's skull to create more space for the swollen brain. It can make the difference between life and death but there was no point in taking half the woman's skull off if she was going to be left wrecked anyway. 'She might make quite a good recovery.'

'Really?' I asked.

'Well, she might do ...'

I said nothing for a while and looked sadly at the scan. I noticed that she was almost exactly my own age.

'It's not my operating day today,' I remarked eventually. 'I suppose we ought to give her the benefit of the doubt.' I said I would see if I could arrange for one of my colleagues to do the operation and made a few phone calls. I settled back to the paperwork again – the operation required was crude and simple but I would have much preferred to be doing it myself instead of reading reports and dictating endless letters. Like all surgeons all I want to do is operate.

After a while I went up to the theatres to see how my colleague was getting on.

I was puzzled to see that the light in the anaesthetic room next to the theatre, which serves as an ante-chamber to the theatre, was turned off and the room was dark. This was most unusual. I pushed the door open and started with shock on entering – there was a shroud-wrapped corpse on the trolley on which patients lie waiting to be anaesthetized. A sheet was wrapped around the lifeless body and tied with a large knot at the top, so that the head was hidden. It looked like a figure from a medieval painting of the Dance of Death.

Feeling very uneasy I walked past the inexplicable corpse and put my head through the theatre doors where my colleague and the nurses and anaesthetists were starting the operation on the woman with ADEM. I found myself in a quandary – had they had a death on the table? Where did the dead body come from? Deaths during an operation are very rare – I have only experienced that most horrible of surgical disasters four times in my career – and the atmosphere in the theatre afterwards was always grim and sombre. The nurses would sometimes be in tears, and I myself would be close to tears, especially if the patient that had died was a child. Yet my colleague and his team all seemed quite cheerful and, I felt, were silently laughing at me. I felt too embarrassed to ask why there was a dead body in the anaesthetic room – if there had been a death on the table I did not want to hurt their feelings by drawing attention to it. So instead I asked him how he was going to carry out the decompressive craniectomy.

He was standing at the patient's head, brilliantly illuminated by the operating lights. Her hair had been shaved and he was now painting her bare and depersonalized head with brown antiseptic iodine.

'Oh, a big bi-frontal craniotomy,' he said. He was going to saw off the front of the woman's skull in order to allow the woman's brain to swell out of its bony confines. Afterwards one simply stitches the scalp closed and, if the patient survives, one can replace the bone taken off the skull at a later date once the swelling has resolved.

I was feeling very uncomfortable, almost frightened, as I talked. I could feel the sinister presence behind me in the darkened anaesthetic room of the shrouded corpse only a few inches away. I asked him what he would do with the falx, the sheet of meninges that separates the two cerebral hemispheres and which might damage the woman's brain as it swelled out of the opened skull.

'Divide it, having sacrificed the saggital sinus anteriorly,' he replied. We continued in this technical vein for a while until eventually I summoned up my courage to ask about the corpse.

'Oh,' he said with a laugh, and the rest of the theatre team laughed with him. 'You noticed! It's just an organ donor – a brain dead head injury from the ITU. Rather, what's left of him. That cyclist from two nights ago. He didn't make it despite surgery. Probably a good thing. The transplant team did a snatch last night. Heart, lung, liver and kidneys – they took the lot, all in good nick. They were delighted. They finished later than usual and the porters were changing shift so they haven't got round to taking him away yet.'

11

EPENDYMOMA

n. a cerebral tumour derived from the non-nervous cells lining the cavities of the ventricles of the brain.

There was little operating to do but much paperwork waiting for me in my office when I returned from a two-week trip to China to visit my eldest daughter Sarah, who was working in Beijing. The paperwork had been organized into a series of threatening piles by Gail – I suspect with a certain vengeful glee, since we are in a state of constant warfare trying to offload paperwork into each other's offices. The many emails from the hospital management I deleted without reading. Among the letters was one from a doctor in a hospital in Lincolnshire asking for my advice about one of my patients – a young woman I had operated upon three times over the preceding ten years for a brain tumour called an ependymoma that kept on recurring, becoming more aggressive and malignant on each occasion. She had had all the possible radiotherapy and chemotherapy and had now been admitted to her local hospital as a terminal case, with severe headaches from yet further recurrence of the tumour. Would I have a look at the latest scan, the doctor asked, and see if anything more could be done, since the family were finding it difficult to accept that the girl was reaching the end of her life.

I had got to know Helen well over the years and had

come to like her greatly. Perhaps that was a mistake. She was always charming and seemed to cope remarkably well with her illness although sometimes I wondered if that was because she was simply unrealistic about her prospects. Denial is not always a bad thing. Her family were devoted to her and, while thanking me effusively whenever I had seen them, would look at me with such an intensity of hope and desperation that their eyes felt like nail guns fixing me to the wall.

'The family have been told by a neurosurgeon in another hospital,' the doctor in Lincolnshire had added in his letter, 'that if the tumour is operated on again he can then treat it with photo-dynamic therapy and they are desperate that you should do that so she can then have the treatment.' Helen's latest scan accompanied the letter on a CD and after the usual delays and swearing I managed to view it on my office computer. It showed extensive recurrence of the tumour in the right temporal lobe of her brain – an area on which in theory I could operate again but where further surgery, if successful, would only provide a few extra weeks or months of life at best.

It was clear that her family had had their hopes raised falsely – photo-dynamic therapy had been shown some time ago to be of little use and I was angry that it had been suggested. It seemed unlikely, however, that her family would accept that nothing further could be done and I knew that they would want anything done, even if it only added a few weeks to her life. With little enthusiasm I telephoned my registrar and asked him to arrange for the patient to be transferred to our hospital.

Over the course of the day, into the evening, I received a series of phone calls and text messages about the patient and the seemingly insuperable problems of transferring her from one hospital to another. Helen was said to be unconscious, and would now need to be transferred on a ventilator and would therefore need an ITU bed on arrival. We had no ITU beds. I suggested the local doctors try their neighbouring neurosurgical unit though I knew that my colleagues there would not be impressed by my plan of operating on such a hopeless case, but then, I told myself, they did not know the family. And then I was told she was better and would no longer need an ITU bed. I then rang my registrar who said that we had ward beds and that we would be able to admit her. At ten o'clock in the evening the hospital in Lincolnshire rang me to say that they had been told by my hospital's bed manager that we did not have any beds.

In a state of increasing irritation I drove there myself to find a bed and the nurse in charge of admissions. I found her – a highly competent nurse, with whom I have worked for years – by the nurse's station.

'Why can't we admit the patient from Lincolnshire?' I asked.

'I'm sorry Mr Marsh, we are waiting for the London ambulance to collect another patient and we cannot accept the new patient until the bed is empty,' she replied.

'But she's coming from over a hundred miles away!' I said, almost shouting. 'She'll arrive in the middle of the bloody night if you insist on waiting for the ambulance to collect the other patient first.'

The nurse looked anxiously at me, and I worried she was going to burst into tears.

'Look, just tell them to send her,' I said, struggling to speak more gently. 'If there are problems say it's all my fault and that I insisted ...'

She nodded her head and said nothing, obviously unhappy about my asking her to break some management protocol about how to admit patients. I felt unable to ask her what she was going to do, reluctant to upset her any further. I turned away and went home. In the past, this would never have happened – an extra bed would always have been found, nor would anybody have questioned my instructions.

Helen finally arrived in the middle of the night, though when I went into work in the morning nobody knew to which ward she had been admitted and I went to the morning meeting without having seen her. At the meeting I told the on-call registrar to put the patient's brain scan up. I gave a brief summary of Helen's history.

'Why do you think I am operating on this hopeless case?' I asked the juniors. No one volunteered a reply, so I explained about her family and how they found it impossible to accept that nothing more could be done.

With slowly progressive cancers it can be very difficult to know when to stop. The patients and their families become unrealistic and start to think that they can go on being treated forever, that the end will never come, that death can be forever postponed. They cling to life. I told the meeting about a similar problem some years ago of a three-year-old child, an only child from IVF treatment. I had operated for a malignant ependymoma and he was fine, and had radiotherapy afterwards. When it recurred – which ependymomas always do – two years later, I operated again and it recurred again, deep in the brain, soon afterwards. I refused to operate another time – it seemed pointless. The conversation with his parents was terrible: they wouldn't accept what I said and they found a neurosurgeon elsewhere who operated three times over the next year and the boy still died. His parents then tried to sue me for negligence. It was one of the reasons I stopped doing paediatrics. Love, I reminded my trainees, can be very selfish.

'Is that why you're operating on this case? Worried that you might be sued?' somebody asked.

In fact I was not worried about being sued, but I was worried that I was being a coward, or maybe just lazy. Perhaps

I was going to operate because I couldn't face confronting the family and telling them it was time for Helen to die. Besides, the cancer specialists think it's a big success if the latest expensive new drug keeps a patient alive for an extra few months.

'What's the photo-dynamic therapy?' somebody else asked.

'Shining laser light on the tumour,' my colleague Francis explained. 'It only penetrates one millimetre and has been shown to be pretty useless. To recommend it now is highly dubious. I think you're daft,' he added, looking at me. 'This will be her fourth op, she's had radiotherapy, the tumour will grow back within weeks – she's at high risk of the bone flap getting infected and then you'll have to remove it and leave her with a big hole under her scalp and she'll die slowly and miserably with a fungus.'

I couldn't deny that this might happen. I turned to the row of registrars sitting at the back of the room and asked if any of them had seen a *fungus cerebri*.

It seemed nobody had and I hoped that none of them ever would. I have only seen it once and that was in Ukraine. If you've had to take the bone flap out after operating on a malignant tumour because the flap's got infected and you don't replace it the patient will then die slowly as the tumour recurs since the tumour can expand outwards through the defect in the skull under the scalp. The patient looks like a Star Trek alien with an extra bit of brain. You don't die quickly from raised intracranial pressure as you do when the skull is intact.

'Can't you put a metal plate back in?' one of the juniors asked.

'It will probably just get infected in turn,' I said.

'If the bone flap gets infected why don't you just leave it in?' he asked.

'And have pus pouring out of the patient's head? Maybe

if the patient was at home, but you can't very well leave an open infection on a hospital ward,' Francis said. 'Well, I hope you get away with it, but I still think you're daft. Just say "No".'

I operated later that morning. I found a sad tangle of tumour, dying brain and blood vessels and could achieve next to nothing. As I assisted my registrar stitching Helen's fragile scalp back together I bitterly regretted my weakness in agreeing to operate. These thoughts were interrupted by the anaesthetist.

'One of the managers was round here earlier,' he said. 'She was very angry that you were admitting patients when there wasn't a bed and said you shouldn't be operating on this case anyway.'

'That's none of her bloody business,' I growled. 'I make the clinical decisions here, she doesn't. Maybe she would like to go and talk to the family and tell them it's time for Helen to die, or that we haven't got any beds ...'

My hands were starting to shake with anger and I had to make a conscious effort to calm down and get on with the operation.

When her scalp was closed my registrar and I stood back and looked at the girl's head.

'It's not going to heal very well, is it?' he commented, young enough still to enjoy the drama and tragedy of medicine.

'You haven't seen a fungus,' I replied.

Afterwards I sat down with the family in one of the little rooms off the ward dedicated to 'breaking bad news'. I did my best to deprive them of all hope, which rather contradicted any reason for operating in the first place, so I was not pleased with myself. I told them that I didn't think the operation was going to make any useful difference and it was only a matter of time before Helen died.

'I know you feel unhappy about operating again,' her brother said to me after I had finished, 'but we want you to know how grateful we are. None of the other doctors would listen to us. She knows she's going to die. She just wants a little longer, that's all.'

As he spoke, I could see through the window that it was a fine spring morning, and even the dull hospital courtyard outside seemed a little hopeful.

'Well, if we're lucky she might get a few extra months,' I said, trying to soften the blow, already regretting how I had spoken to them a few minutes earlier, failing to find a balance between hope and reality.

I left them in the little room, their knees squeezed together as the four of them sat on the small sofa and wondered, yet again, as I walked away down the dark hospital corridor, at the way we cling so tightly to life and how there would be so much less suffering if we did not. Life without hope is hopelessly difficult but at the end hope can so easily make fools of us all.

The next day was even worse. None of us felt able to make our usual sardonic jokes at the morning meeting. The first case was a man who had died as a result of an entirely avoidable delay in his being transferred to our unit; another was a young woman who had become brain dead after a haemorrhage. We looked glumly at her brain scan.

'That's a dead brain,' one of my colleagues explained to the juniors. 'Brain looks like ground glass.'

The last case was an eight-year-old who had tried to hang himself and had suffered hypoxic brain damage.

'Can we have some rather less depressing cases, please?' someone asked, but there were none and the meeting came to an end.

As I was leaving, one of the neurologists came down the

corridor looking for me. He was sporting a three-piece suit – a rarity for hospital consultants in the modern age – but instead of looking his usual jovial and positive self he appeared a little hesitant.

'Can I ask you to see a patient?' he asked.

'Of course,' I replied enthusiastically, always keen to find more patients for surgery and hoping for a benign tumour but a little worried by his expression.

'The scans are on PACS,' he said and we went back into the viewing room where his registrar summoned up a brain scan on PACS, the digital X-ray system, on one of the computers.

'She's only thirty-two, I'm afraid,' the neurologist explained.

'Oh dear,' I said. The scan showed a large and unmistakably malignant tumour at the front of her brain.

'It seems to be a bad week.'

We walked to the day ward where the patient was lying behind curtains on a bed. She had had the scan twenty minutes earlier and the neurologist had only just told her – in broad terms – what it showed. A young mother, with two children, she had been suffering from headaches for a few weeks. Her husband was sitting by the bedside. It was obvious that they had both been crying.

I sat down on her bed and did my best to explain what treatment she would need. I tried to give her some hope but could not pretend that she was going to be cured. With these terrible conversations, especially when the bad news is being broken so suddenly, all doctors know that patients will only take in a small part of what they are told. I sent her home on steroids – which would get the headaches quickly better – and arranged to operate next Monday, promising her and her distraught husband that I would explain everything again when she was admitted the evening before the operation. It does not feel very good to tell somebody, in effect, that they have an incurable brain tumour and then tell them to go home, but there was nothing else to be done.

Next morning I showed her brain scan to the juniors at the morning meeting. It appeared on the wall in front of us in black and white.

I told them the story and asked David, one of the younger trainees, to imagine that he had been asked to see her after the scan, as I had been the previous day. I asked him what he would tell her.

David, normally full of confidence and enthusiasm, was silent.

'Come on,' I said 'You've got to say something to her. You must have had to do this before.'

'Er, well ...' he fumbled for the words, 'I'd tell her that there was an abnormality of the scan with, er, mass effect ...'

'What the fuck will that mean to her?' I said.

'I'd tell her she needed to have an operation so that we could find out what it is ...'

'But you're lying. We know what it is, don't we? It's a highly malignant tumour with an awful prognosis! You're scared of telling her! But she'll know it's bad just from the way you look at her. If it was a benign tumour you'd be all smiles, wouldn't you. So what are you going to tell her?'

David said nothing and there was an awkward silence in the darkness of the X-ray viewing room.

'Well, it's very difficult,' I said in a gentler tone. 'That's why I was asking you.' When I have had to break bad news I never know whether I have done it well or not. The patients aren't going to ring me up afterwards and say 'Mr Marsh, I really liked the way you told me that I was going to die,' or 'Mr Marsh, you were crap'. You can only hope that you haven't made too much of a mess of it.

Surgeons must always tell the truth but rarely, if ever, deprive patients of all hope. It can be very difficult to find the balance between optimism and realism. There are degrees of malignancy with tumours and you never know what will happen to the individual patient in front of you – there are always a few long-term survivors – not miracles but statistical outliers. So I tell my patients that if they are lucky they might live for many years, and if unlucky it might be much less. I tell them that when the tumour recurs it might be possible to treat them again and, although to some extent it is clutching at straws, you can always hope that some new treatment will be found. Besides, most patients and their families will re-search their disease online and the paternalistic white lies of the past will no longer be believed. Nevertheless, sooner or later, most of the patients, like Helen, will reach the point of no return. It is often very difficult for both doctor and patient to admit that it has been reached. The juniors in the morning meeting listened in the dark in respectful silence as I tried to explain all this, but I do not know if they really understood.

After the meeting I went back onto the ward to see Helen.

Mary, the ward sister came up to me.

'The family are completely unrealistic,' she pointed to the door of the side-room where Helen was lying. 'She's obviously dying but they just won't accept it.'

'What's the plan?' I asked.

'The family won't let us treat her as a terminal case with decent painkillers so we're trying to get community services and the GP organized and get her home.'

'And the wound?' I dreaded the answer.

'It looks like it's going to break down any moment.'

I took a deep breath and went in to the side-room. To my relief the family were not present. Helen was lying on her side facing the window so I walked round the bed and squatted down beside her. She looked at me, her eyes large and dark, and slowly smiled. The right side of her head was swollen but covered by a wound dressing. I saw little point in uncovering it so I left it undisturbed, and spared myself a sight any surgeon hates – a once-neat incision for which he has been responsible breaking down and becoming an ugly, gaping wound.

'Hello Mr Marsh,' she said.

It was difficult to know what to say.

'How are you?' I offered.

'Getting better. Head hurts a bit.' She spoke slowly, her speech a little slurred by her left-sided paralysis. 'Thank you for operating again.'

'We'll get you home as soon as we can,' I said. 'Any questions?' I resisted the temptation to get up and walk to the door as I asked the question – an unconscious trick all doctors have to fight against when faced by a painful conversation. Helen said nothing so I left and went to the operating theatres.

12

GLIOBLASTOMA

n. the most aggressive type of brain tumour derived from non-nervous tissue.

I have little direct contact with death in my work despite its constant presence. Death has become sanitized and remote. Most of the patients who die under my care in the hospital have hopeless head injuries or cerebral haemorrhages. They are admitted in coma and die in coma in the warehouse space of the Intensive Care Unit after being kept alive for a while by ventilators. Death comes simply and quietly when they are diagnosed to be brain dead and the ventilator is switched off. There are no dying words or last breaths – a few switches are turned and the ventilator then stops its rhythmic sighing. If the cardiac monitor leads have been left attached – usually they are not – you can watch the heart on the ECG monitor – a graphic line in LED red that rises and falls with each heart beat – become increasingly irregular as the dying heart, starved of oxygen, struggles to survive. After a few minutes, in complete silence, it comes to a stop and the trace becomes a flat line. The nurses then remove the many tubes and wires connected to the now lifeless body and after a while two porters bring a trolley with a shallow box beneath it camouflaged by a blanket and wheel the body away to the mortuary. If the patient's organs are to be used for donation the ventilator will be kept running after the brain's death has been certified and the body will be taken to the operating theatres – usually at night. The organs are removed and only then is the ventilator switched off and the camouflaged trolley will come to take the corpse away.

The patients I treat with fatal brain tumours will die at home or in hospices or in their local hospital. Very occasionally one of these patients of mine with a brain tumour will die under my care while still in the hospital but they will be in coma, since they are dying because their brain is dying. If there are any discussions about death or dying it is with the family and not with the patient. I rarely have to confront death face to face, but occasionally I am caught out.

When I was a junior doctor it was very different. I was closely involved with death and with dying patients on a daily basis. In my first year as a doctor, working as a houseman at the bottom of the medical hierarchy, I would often be summoned, usually out of bed in the early hours, to certify the death of a patient. I would walk along the empty, anonymous corridors of the hospital, young and healthy and wearing a doctor's white coat, to enter a dark ward and be directed by the nurses to a bed around which the curtains had been drawn. I would be aware of the other patients, usually old and frail, lying in the neighbouring beds, probably awake and terrified in the dark, probably thinking of their own fate, desperate to recover and escape the hospital.

The dead patient behind the curtain, faintly lit by a dim bedside light, would look like all dead hospital patients. They would usually be elderly, in a hospital gown, as anonymous as Everyman with a pinched, wax-yellow face and sunken cheeks, purple blotches on the limbs and utterly still. I would open the gown and place my stethoscope over the heart to confirm that there was no heartbeat and then open the eyelids and shine a small pen torch into the dead eyes to check that the pupils were 'fixed and dilated' – that they were dull and black, as large as saucers, and did not constrict in reaction to the torch's light. I would then go to the nurses' station and write in the notes 'Certified dead', or words to that effect, and sometimes I would add RIP. I would sign this and then go back to bed in the little on-call room. Most of the patients I certified in this way were not known to me – at night I would be covering wards with patients who belonged to different firms from the one for which I worked during the day. This was many years ago, when post-mortems were still common practice. It was traditional to attend the post-mortems of the patients who had died on the wards for which you were responsible in the daytime, whom you had cared for in their final illness and whom you had got to know. But I hated post-mortems and usually tried to avoid them. My detachment had its limits.

As a casualty officer – the next job I had after my year as a houseman and then as a trainee in general surgery – I saw death in more dramatic and violent forms. I remember patients dying from heart attacks – or 'arresting' – in front of me. I remember working all night trying, and failing, to save one man, fully awake and suffering horribly, who looked into my eyes as he bled to death from oesophageal varices. I have seen people die from gunshot wounds, or crushed and broken in car crashes, or from electrocution, heart attacks, asthma, and all manner of cancers, some of them quite repulsive.

And then there were the 'BIDs' – people brought in dead by the ambulance men. As the casualty officer I would have to certify death in some poor soul who had collapsed and died in the street. On these occasions I would find the corpse fully dressed on a trolley, and having to undo their clothing to place my stethoscope on their heart was a profoundly different experience from certifying death in the hospital inpatients in their anonymous white gowns. I felt that I was assaulting them, and I wanted to apologize to them as I unbuttoned their clothes, even though they were dead. It is remarkable how much difference clothing makes.

I was driving out of London on a Friday afternoon, about to take a few days' leave and have a brief holiday with my wife. It had been a cold winter and I was admiring the way in which the branches of the trees beside the motorway were elegantly outlined with snow when my mobile rang. Checking that there were no police cars in sight, I answered it. I could not catch what was said.

'Who?' I asked.

I could not make out the name but the voice on the phone said. 'We've just admitted your patient David H— from his home.'

'Oh,' I said and pulled onto the hard shoulder.

'He's got a progressive hemiparesis and had become increasingly drowsy but he's better with steroids – quite sharp and witty again.'

I remembered David very well. I had first operated on him twelve years earlier for a particular species of tumour called a low grade astrocytoma of the right temporal lobe. These are tumours within the brain itself that at first grow slowly, initially causing the occasional epileptic fit but which eventually undergo malignant transformation and become 'high grade' tumours known as glioblastomas and ultimately prove fatal. This can take many years and it is impossible to predict for any particular patient how long they have to live. A few tumours, if they are small enough, can be cured by surgery. Most of the patients are young adults who must learn to live with a slow death sentence. With these patients it is especially difficult to know how to explain their diagnosis to them. If you get the balance between optimism and realism wrong – as I sometimes do however hard I try – you can either condemn the patient to live in hopeless despair for whatever time they have left or end up being accused of dishonesty or incompetence when the tumour turns malignant and the patient realizes that they are going to die. David, however, had always made it clear to me that he wanted to know the truth, however grim and uncertain.

He was in his early thirties when he had his first fit and the tumour was discovered, a successful management consultant, over six foot tall and a keen cyclist and runner. He was married with young children. He was a person of great charm and determination, who managed to turn everything into a joke, and continued to make jokes even when I had opened his head and was removing his tumour under local anaesthetic while he was wide awake. We had both hoped that he would be one of the fortunate few cured by surgery but after three years the follow-up scans showed that the tumour was back. I remember very clearly telling him this as he sat across from me in the outpatient room, that this meant that eventually the tumour would kill him. I could see the tears forming in his eyes as I talked, but he swallowed hard and looked straight ahead for a few moments and we then discussed what further treatment might buy him some more time. Over the next few years I operated twice again, and, with radiotherapy and chemotherapy, he had managed to go on working and leading a normal life until only recently. Compared to many other people with these tumours he had, as doctors put it, 'done very well'. I had got to know him, and his wife, better than some of my patients over those years and I was humbled by how they bore his illness, and of how they managed to be so practical and determined about it.

'I don't think there's anything else to be done,' the doctor calling me said down the phone, 'but he would like you to look at his scans. He has great faith in you. I've already shown them to one of the neurosurgeons here and he wasn't keen.'

'I'm leaving the country tomorrow morning for a few days,' I said. 'Send the scans to me electronically and I'll have a look at them next week.'

'Of course,' she replied, 'I'll do that. Thanks.'

Snow was now falling. As I pulled back onto the motorway and continued on my way I found myself engaged in a painful internal dialogue. By chance, I was only a short distance away from the hospital to which David had been admitted and it would be a minor detour to go and see him in person.

'I really don't want to go and tell him he is going to die,' I said to myself, 'I don't want to spoil a nice weekend away with my wife.' But I felt a sort of dragging sensation deep inside myself.

'Ultimately, if I was dying,' I heard myself saying, 'wouldn't I appreciate a visit from the surgeon in whom I have put my hope for so many years? ... But I really don't want to tell him it is time to die ...'

Angrily, reluctantly, I took the next turning off the motorway and drove to the hospital. It rose like a monolith out of a huge surrounding car park. I walked unhappily along the endless long central corridor inside. It seemed to go on for miles and miles but perhaps this was the effect of my dread at going to talk to my dying patient. I experienced once again my visceral hatred of hospitals and their dull, indifferent architecture within the walls of which so much human suffering must be acted out.

At least the lifts, as I ascended to the fifth floor, didn't tell me to wash my hands as they do in my own hospital, but the voice telling me when the doors were opening and closing sounded even more irritating than usual.

I finally walked onto the ward. I found David standing by the nurses' station, in pyjamas, towering above a little group of nurses who were propping him upright. He was leaning a little to one side because of his left-sided weakness.

The doctor who had phoned me was standing next to him and came towards me.

'They all think I'm a magician! I ring you up and within fifteen minutes you appear!'

I walked up to David who laughed with amazement at my sudden arrival.

'You again!' he said.

'Yes,' I said. 'I'll go and look at the scans.' I was taken to a nearby computer.

I had not met the doctor caring for him before although we had exchanged letters about David. It was immediately obvious that she was deeply sympathetic.

'I look after all the patients with the low grade gliomas,' she said with a slight grimace. 'Motor Neurone Disease and MS are easy by comparison. The patients with low grade gliomas are all young, with young children and all I can say is go away and die ... My children are the same age as David's, go to the same school. It's difficult not to get involved, not to get emotional.'

I looked at the scan on the computer. It showed that the tumour, which was now cancerous, was burrowing deep into his brain. The fact that the tumour was on the right side of his brain meant, as had been the case with Helen, that his intellect and understanding were still largely intact.

'Well I could operate,' I said, 'but it probably would not buy him much more time ... a few months at best. It would be prolonging dying, not living. It would waste what little time he has left with false hope and would not be without risk. He's always made it clear to me that he wanted to know the truth.' I thought of the other patients I had re-operated on in the past in similar circumstances, such as Helen, who could not bear to face the truth, and how I had usually regretted it. But it is so very difficult to tell your patient that there is nothing more that can be done, that there is no hope left, that it is time to die. And then there is always the fear that you might be wrong, that maybe the patient is right to hope against hope, to hope for a miracle, and maybe you should operate just one more time. It can become a sort of *folie à deux*, where both doctor and patient cannot bear reality.

While I was looking at the scans David had been guided back to the single room to which he had been admitted the previous day, unconscious and half-paralysed, before the high dose steroid drugs temporarily brought him back to life.

I walked into the room, where his wife and two nurses were standing at the end of the bed. The afternoon light was fading and the room was dark as the electric lights had not yet been switched on. Through the window I could see the sombre day outside, and the hospital car park a few storeys below us, and beyond that a line of trees and houses, with snow falling but not settling on the ground.

David was lying on his back and turned with an effort towards me as I came in. I stood a little nervously above him.

'I've been looking at the scans.' I paused. 'I always told you I'd tell you the truth.'

I noticed that he was not looking at me as I spoke and I realized that I was on his left, hemianopic side. He probably could not see me as the right side of his brain was no longer working so I walked round the bed and, with my knees cracking, knelt down beside him. To stand over your dying patient would be as inhuman as the long hospital corridors. We looked into each other's eyes for a moment.

'I could operate again,' I said slowly, having to force the words out, 'but it would only get you an extra month or two at best ... I have sometimes operated on people in your situation ... I usually regretted it.'

David started to reply, talking equally slowly.

'I realized things did not look good. There were ... various things I needed to organize but I ... have ... done that all now ...'

I have learned over the years that when 'breaking bad news' as it is called, it is probably best to speak as little as possible. These conversations, by their very nature, are slow and painful and I must overcome my urge to talk and talk to fill the sad silence. I hope I do these things better than I did in the past, but I struggled as David looked at me and I found it difficult not to talk too much. I said that if he was a member of my own family I would not want him to have any more treatment.

'Well,' I said eventually, getting control of myself, 'I suppose I've kept you going for a good few years ...'

In the past he had been a competitive cyclist and runner and he had large, muscular arms. I felt awkward as I shyly reached out to hold his big, masculine hand.

'It's been an honour to look after you,' I said and stood up to leave.

'It's a bit inappropriate but all I can say is good luck,' I added, unable to say goodbye, since we both knew it would be for the last time.

I stood up – his wife came towards me, her eyes full of tears.

I buried my face in her shoulder, holding her fiercely for a few seconds and then left the room. His doctor followed me.

'Thank you so much for coming. It will make everything much easier. We'll get him home and organize palliative care,' she said.

I waved my hands despairingly in the air and walked away, imitating the staggering walk of a drunk, drunk on too much emotion.

'I'm happy,' I called back to her as I walked down the corridor. 'It was good, so to speak, to have that conversation.'

Will I be so brave and dignified when my time comes? I asked myself as I walked out into the grim black asphalt car park. The snow was still falling and I thought yet again of how I hate hospitals.

I drove away in a turmoil of confused emotions. I quickly became stuck in the rush-hour traffic, and furiously cursed the cars and their drivers as though it was their fault that this good and noble man should die and leave his wife a widow and his young children fatherless. I shouted and cried and stupidly hit the steering wheel with my fists. And I felt shame, not at my failure to save his life – his treatment had been as good as it could be – but at my loss of professional detachment and what felt like the vulgarity of my distress compared to his composure and his family's suffering, to which I could only bear impotent witness.

13

INFARCT

n. a small localized area of dead tissue caused by an inadequate blood supply.

On one of my regular trips to the neurosurgical department in America where I have an honorary teaching post I delivered a lecture entitled 'All My Worst Mistakes'. It had been inspired by Daniel Kahneman's book *Thinking, Fast and Slow*, a brilliant account, published in 2011, of the limits of human reason, and of the way in which we all suffer from what psychologists call 'cognitive biases'. I found it consoling, when thinking about some of the mistakes I have made in my career, to learn that errors of judgement and the propensity to make mistakes are, so to speak, built in to the human brain. I felt that perhaps I could be forgiven for some of the mistakes I have made over the years.

Everybody accepts that we all make mistakes, and that we learn from them. The problem is that when doctors such as myself make mistakes the consequences can be catastrophic for our patients. Most surgeons – there are always a few exceptions – feel a deep sense of shame when their patients suffer or die as a result of their efforts, a sense of shame which is made all the worse if litigation follows. Surgeons find it difficult to admit to making mistakes, to themselves as well as to others, and there are all manner of ways in which they disguise their errors and try to put the blame elsewhere. Yet as I approach the end of my career I feel an increasing obligation to bear witness to past mistakes I have made, in the hope that my trainees will learn how not to make the same mistakes themselves.

Inspired by Kahneman's book I set out to remember all my worst mistakes. For several months, each morning, I would lie in bed before getting up to head off for my daily run round the local park, thinking over my career. It was a painful experience. The more I thought about the past the more mistakes rose to the surface, like poisonous methane stirred up from a stagnant pond. Many had been submerged for years. I also found that if I did not immediately write them down I would often forget them all over again. Some, of course, I have never forgotten and that has usually been when the consequences for myself had been especially unpleasant.

When I delivered my lecture to my American colleagues, it was met by a stunned silence and no questions were asked. For all I know they may have been stunned not so much by my reckless honesty as by my incompetence.

Surgeons are supposed to talk about their mistakes at regular 'Morbidity and Mortality' meetings, where avoidable mistakes are discussed and lessons learned, but the ones I have attended, both in America and in my own department are usually rather tame affairs, with the doctors present reluctant to criticize each other in public. Although there is much talk of the need for doctors to work in a 'blame-free' culture it is very difficult in practice to achieve this. Only if the doctors hate each other, or are locked in furious competition (usually over private practice, which means money), will they criticize each other more openly, and even then it is more often behind each other's backs.

*

One of the mistakes I discussed in my lecture, and which I had not forgotten, involved a young man who had been admitted to the old hospital, shortly before it closed. My registrar – in fact one of the American trainees who are sent from their department in Seattle to work in London in my hospital for a year as part of their training – came to find me and asked me to look at a scan.

We walked from my office to the X-ray viewing room. This was before the X-ray system had been computerized and the patients' brain scans were all on large sheets of film. They were kept on chrome and steel frames on which the films were hung like washing on a line, each frame on roller bearings so that the frames could be pulled out smoothly, one by one. The system was like an antique Rolls Royce – old-fashioned but beautifully engineered. Provided that you had highly efficient X-ray secretaries – which we did – the system was completely reliable and quite unlike the computers that now dominate my working life. My registrar pulled out some scans in front of me.

'He's a thirty-two-year-old man at St Richard's – apparently he's become paralysed down the left side,' he told me.

The scan showed a large dark area on the right side of the man's brain.

To the man with a hammer, it is said, all things look like nails. When brain surgeons look at brain scans they see things that they think require surgery and I am, alas, no exception. I looked at the scan quickly – I was already late for my outpatient clinic.

I agreed with my registrar that it was probably a tumour but one which was impossible to remove. All that could be done would be a biopsy operation where a small part of the tumour would be removed and sent for analysis. I told him to bring the patient over to our hospital for this to be done. In retrospect I was careless – I should have asked more questions about the history and if I had been given the right information, which admittedly I might not have been since it was all at second hand, I would probably have looked more critically at the scans or asked for my neuroradiologist's opinion.

So the young man was transferred to the neurosurgical unit. My registrar duly carried out the biopsy operation – a minor and relatively safe operation done through a half inch hole drilled in the skull that took less than an hour. The analysis came back that the abnormality was not a tumour but an infarct – he had a suffered a stroke, unusual in a man his age, but not unheard of. In retrospect it was rather obvious that this was what the scan had shown and I had misinterpreted it. I was embarrassed but not especially troubled – it did not seem too terrible a mistake to make and a stroke seemed better than a malignant tumour. The patient was transferred back to the local hospital to be investigated for the cause of his stroke. I thought nothing more about it.

Two years later I received a copy of a long letter, written in a shaky, elderly hand, from the man's father. The letter had been sent to the hospital and passed on to me for my comments by the Complaints Office, recently renamed by the new chief executive as the 'Complaints and Improvements Department'. The letter accused me of being responsible for the death of his son who had died several months after being transferred back to the local hospital. His father was certain he had died because of the operation.

I invariably become very anxious when I receive letters of complaint. Every day I will make several dozen decisions that, if they are wrong, can have terrible consequences. My patients desperately need to believe in me, and I need to believe in myself as well. The delicate tight-rope walking act of brain surgery is made all the worse by the constant pressure to get patients in and out of hospital as quickly as possible. When I receive one of these letters, or one from a solicitor announcing the intention of a patient of mine to sue me, I am forced to see the great distance beneath the rope on which I am balancing and the ground below. I feel as though I am about to fall into a frightening world where the usual roles are reversed – a world in which I am powerless and at the mercy of patients who are guided by suave, invulnerable lawyers who, to confuse me even further, are dressed in respectable suits just as I am and speak in the same self-confident tones. I feel that I have lost all the credibility and authority that I wear like armour when I do my round on the wards or when I open a patient's head in the operating theatre.

I called in the dead man's notes and learned that he had died from a further stroke caused by a disease affecting the blood vessels in his brain that had resulted from the first stroke, which I had misinterpreted as a tumour. The biopsy operation was unnecessary and unfortunate but irrelevant. I explained, apologized and defended myself in a series of letters, which the hospital management rewrote in the third person and sent off to the father with the chief executive's signature. The father was not satisfied and demanded a Complaints meeting which duly took place some months later. A smartly dressed middle-aged woman from the Complaints and Improvements Department, who I had never met before and who obviously knew nothing about the details of the case, chaired the meeting. The dead man's elderly parents sat opposite me, glaring with hatred and anger, convinced my incompetence had killed their son.

I spoke to the parents, unnerved and frightened by their anger, and became quite upset. I tried to apologize but also to explain forcefully why the operation, although a mistake, had nothing to do with their son's death. I had never had to attend such a meeting before and I don't doubt that I made a mess of it. The Complaints and Improvements Manager interrupted me and told me I should listen to what the patient's father had to say.

I therefore had to sit for what felt like a very long time while the bereaved man poured out his grief and anger. I was told afterwards by another manager present at the meeting that the lady from Complaints and Improvements was silently weeping as the old man described his suffering, for which I was held to be uniquely responsible. I learned later that the day of the meeting was the second anniversary of his son's death and he had been to visit his grave at the local cemetery that morning. The Complaints and Improvements manager eventually dismissed me and I left the room feeling very shaken.

I thought that would be the end of the matter but a few weeks later the chief executive of the hospital Trust rang me on my mobile, entirely out of the blue, just a few days before Christmas. He was a new appointment, recently parachuted in by the Department of Health because of the Trust's parlous financial situation. His predecessor had been suddenly and ignominiously dismissed. I had met this new chief executive briefly when he started. As with all NHS chief executives in my experience (I have now got through eight) they do the rounds of the hospital departments when they are appointed and then one never sees them again, unless one is in trouble, that is. This is called Management, I believe.

'I'm giving you advance warning of a meeting you are to have with me in the New Year,' he announced.

'But what's it about?' I asked, immediately anxious.

'That will have to wait until the meeting.'

'For Christ's sake – why are you ringing me up now then?'

'To give you advance warning.'

I felt frightened and confused and could only assume this was the desired effect of the phone call.

'What am I supposed to make of that? Advance warning about what? I've just about had enough of working here,' I said pathetically. 'I feel like resigning.'

'Oh we can't have that,' he replied.

'Well, tell me what the problem is then!' I cried.

'It is about a recent Complaints meeting, but it will have to wait until the meeting.'

He refused to tell me any more and the conversation ended.

'Happy Christmas,' I said to my mobile phone.

The meeting was scheduled for early January and I spent much of Christmas brooding over it. I may appear to others to be brave and outspoken but I have a deep fear of authority, even of NHS managers, despite the fact that I have no respect for them. I suppose this fear was ingrained in me by an expensive English private education fifty years ago, as was a simultaneous disdain for mere managers. I was filled with ignominious dread at the thought of being summoned to meet the chief executive.

In the event, a few days before I was to meet him I suffered a haemorrhage into my left eye and had to undergo emergency surgery for a retinal detachment. Perhaps because of my impaired eyesight I then fell down the stairs at home a few weeks later and broke my leg. Once I had recovered from that I then suffered a retinal tear – a lesser problem than a detachment – in my right eye which required further treatment. By the time that I was back at work it seemed that the chief executive had forgotten about me, as had I our conversation on the telephone. I made one of my regular trips to Ukraine and shortly afterwards I was sitting in my office catching up on the paperwork that had accumulated while I had been away.

'You're in trouble again!' Gail shouted through the doorway between our offices. 'The chief executive's secretary telephoned. You've been summoned to a meeting with the chief executive and the director of surgery tomorrow at eight.'

On this occasion I knew well enough what the meeting was going to be about. Two days earlier, after running up the stairs to the second floor on my way to the morning meeting I was taken aback to find that the doors to the female neurosurgical ward had an enormous three-foot by four-foot poster stuck to them. On it there was a huge No Entry sign in ominous red and black with the grim instructions beneath: 'DO NOT ENTER UNLESS YOUR VISIT IS ESSENTIAL. SOME PATIENTS ON THIS WARD HAVE AN INFECTIOUS ILLNESS.'

I turned away in disgust and went in to the X-ray viewing room for the morning meeting. The juniors were discussing the poster. Apparently there had been an outbreak of norovirus on the ward – an unpleasant but usually harmless virus that used to be called winter flu. My colleague Francis marched into the room waving the poster in his hand, which he had clearly pulled off the ward doors.

'How fucking ridiculous can you get?' he shouted. 'Some moron from management has stuck this up on the door to the women's ward. Are we supposed to stop going to see our patients?'

'You're a naughty boy!' I said. 'You'll be in trouble with the management for removing it!'

After the meeting I went down to my office and sent an email to the hospital's director of infection control complaining about the poster. No doubt I was now blamed for its removal.

At eight o'clock next morning, feeling apprehensive and defensive, I made my way along endless corridors to the labyrinth of managerial offices in the heart of the hospital. I passed the doors for the Manager and Deputy Manager for Corporate Strategy, the Interim Manager for Corporate Development, the Director of Governance, the Directors for Business Planning, for Clinical Risk, and for many other departments with names I cannot remember, almost certainly all created as a result of expensive reports by management consultants. The Department for Complaints and Improvements, I noted, had been renamed yet again and was now the Department for Complaints and Compliments.

The chief executive's office was a suite of rooms, with a secretary in the outer room and a large room beyond with a desk at one end and a table with chairs around it at the other. Just like the offices, I thought a little sourly, of all the ex-communist *apparatchiki* and professors I have dealt with in the former Soviet Union. The chief executive, however, was not going to use the bullying and bluster of some of his post-Soviet counterparts and instead welcomed me enthusiastically and offered me coffee. (On the other hand, some of the nicer post-Soviet professors would welcome me in the morning with vodka.) We were joined shortly afterwards by the director of surgery, who said little throughout the meeting, his expression one of irritation and exasperation with me, and of deference to the chief executive. After the usual niceties, the question of the infection control poster came up.

'Just for once,' I said, 'I followed the proper channels. I sent an email to the director of infection control.'

'It caused great offence. You compared the hospital to a concentration camp.'

'Well, I wasn't the one who copied it to everybody in the Trust,' I retorted.

'Did I say that you did?' the chief executive replied in a stern and headmasterly tone.

'I regret saying "concentration camp",' I said with some embarrassment. 'It was silly and a bit over the top. I should have said "prison".'

'But didn't you remove the poster?' asked the chief executive.

'No, I didn't,' I said.

He looked surprised and the room was quiet for a while. I had no intention of sneaking on my colleague.

'And there was a problem with a Complaints meeting last year.'

'Yes, your Trust's Complaints office managed to arrange the meeting on the anniversary of the patient's death.'

'Not "your Trust" Henry,' said the chief executive. '*Our* Trust.'

'The anniversary of a death is the worst possible date for such a meeting. Have you ever come across so-called "Anniversary Reactions"? Grieving relatives are particularly difficult to handle on these occasions.'

'Well, yes. We did have one of those recently, didn't we?' he said, turning to the director of surgery.

'Nor was there any meeting with me beforehand – with your Trust's staff about the basis of the complaint,' I added.

'*Our* Trust,' he corrected me again. 'But it's true that the procedure is that there should have been a meeting beforehand ...'

'Well the procedure wasn't followed but I'm sorry if I handled the meeting badly,' I said. 'But you try sitting opposite the parents of a patient who has died and who are convinced you killed their child. It's even more difficult when the accusation is absurd, even though I did get the diagnosis wrong and he was subjected to an unnecessary operation.'

The chief executive was silent. 'I couldn't do your job,' he said at last.

'Well, I couldn't do yours,' I replied, filled with sudden gratitude for his understanding. I thought of all the government targets, self-serving politicians, tabloid headlines, scandals, deadlines, civil servants, clinical cock-ups, financial crises, patient pressure-groups, trades unions, litigation, complaints and self-important doctors with which an NHS chief executive must deal. The average time for which they serve, not surprisingly, is only four years.

We looked at each other for a few moments.

'But your Communications Office is crap,' I said.

'All I'm asking is that you use your undoubted abilities for Our Trust,' he said.

'We want you to follow established procedures ...' the director of surgical services added, feeling obliged to contribute to the meeting.

After the meeting I made my way back out of the labyrinth and returned to my office. Later that day I emailed the Communications Department my suggestions for a better poster. 'We need your HELP ...' it began, but I never received a reply.

The chief executive left the Trust a few weeks later. He had been re-directed to another Trust with financial difficulties, where no doubt he was to wield the axe again on behalf of the government and the civil servants in the Treasury and Department of Health. I heard a rumour a few months later that he was on sick leave from his new Trust because of stress and, slightly to my surprise, I felt sorry for him.

14

NEUROTMESIS

n. the complete severance of a peripheral nerve. Complete recovery of function is impossible.

On the first day of June, the weather suddenly hot and humid, I cycled to work for the morning meeting. Before setting off I had gone into my small back garden to inspect my three bee hives. The bees were already hard at work, shooting up into the air, probably heading for the flowering lime trees that grow along one side of the local park. As I pedalled to work I thought happily of the honey I would be harvesting later in the summer. I arrived a few minutes late. One of the senior house officers was presenting the cases.

'The first case,' she said 'is a sixty-two-year-old man who works at one of the local hospitals as a security man. He lives on his own and has no next of kin. He was found confused at home. His colleagues had gone round to look for him because he hadn't turned up for work. There were many bruises on his right side and his colleagues said that he had had increasing difficulties with talking over the previous three weeks.'

'Did you see him when he was admitted?' I asked her, knowing that the house officers presenting the cases at the morning meeting will rarely have seen the patients they present because of their short working shifts.

'Well, actually I did,' she said. 'He was dysphasic and had a slight weakness on the right side.'

'So what's the diagnosis going to be?' I asked.

'It's a short history of a progressive neurological deficit. It involves speech,' she replied. 'The bruises on the right side of his body suggest he's falling to the right so he's probably got a progressive problem on the left side of his brain, probably in the frontal lobe.'

'Yes, very good. What sort of problem?'

'Maybe a GBM, or maybe a subdural.'

'Quite right. Let's have a look at the scan.'

As she worked at the computer keyboard the slices of the poor man's brain scan slowly appeared. It showed what was obviously a malignant tumour in the left cerebral hemisphere.

'Looks like a GBM,' somebody said.

There were two medical students that morning in the audience. The SHO turned to them, probably enjoying the fact that there was somebody even lower in the strict medical hierarchy than herself.

'A GBM,' she said in a knowledgeable tone of voice, 'is a glioblastoma multiforme. A very malignant primary brain tumour.'

'These are fatal tumours,' I added for the benefit of the students. 'A man his age with a tumour like this has only a few months – maybe only weeks – to live. If he's treated, which means partial surgical removal and then radiotherapy and chemotherapy afterwards, he'll only live a few months longer at best and he probably won't regain his speech anyway.'

'Well, James,' I said, turning to one of the registrars, 'the SHO has been spot on with the diagnosis. What is the management of this case? And what are the really important points here?'

'He's got a malignant tumour we can't cure,' James

replied. 'He's disabled despite steroids. All we can do is a simple biopsy and refer him for radiotherapy.'

'Yes, but what's really important about the history?'

James hesitated but before he could reply I said that what was important was that he had no next of kin. He'd never get home. He'd never be able to look after himself. He had only a few months of life left whatever we did – and since he had no family he was likely to spend what little time he had left miserably on a geriatric ward somewhere. But I told James he was probably right – it would be easier to get him back to his local hospital if we established the diagnosis formally, so I said that we had better get a biopsy and bounce him off to the oncologists. We could only hope that they'd be sensible and not prolong his suffering by treating him. The fact of the matter was that we already knew the diagnosis from the scan and any operation would be something of a charade.

I pulled out a USB stick from my pocket and walked up to the computer at the front of the viewing room.

'I'll show you all some amazing brain scans from my last trip to Ukraine!' I said but I was interrupted by one of my junior colleagues.

'Excuse me,' he said, 'but the manager responsible for the junior doctors' working hours has very kindly agreed to come and talk to us about the new rota for the registrars and she can't stay beyond nine o'clock since she has another meeting to go to afterwards. She'll be here in a minute.'

I was annoyed that I was not going to be able to show some enormous Ukrainian brain tumours but clearly I had no choice in the matter.

The manager was late, so while we waited for her to arrive I walked round to the operating theatres, to see the only patient for the day's operating. He was waiting in the anaesthetic room, lying on a trolley, a young man with severe sciatica from a simple disc prolapse. I had seen him six months earlier. He was a computer programmer but also a competitive mountain biker and had been training for some kind of national championship when he developed excruciating sciatic pain down his left leg. An MRI scan had shown the cause to be a slipped disc – 'a herniated intervertebral disc causing S1 nerve root compression' in medical terms. His disc prolapse had prevented him from training and he had had to drop out of the mountain biking championships, to his bitter disappointment. He had been very frightened by the prospect of surgery and decided to see if he would get better on his own which, I had told him, often happened if one waited long enough. This had not happened, however, and he had now reluctantly decided to undergo surgery.

'Good morning!' I said, my voice full of surgical reassurance – genuine reassurance since the planned operation was a simple one. Most patients are pleased to see me before their operation, but he looked terrified.

I leant forward and lightly patted his hand. I told him that the operation really was a very simple one. I explained that we always had to warn people of the risks of surgery but promised him that it really was most unlikely that things would go wrong. If I'd had sciatica for six months I would have the op, I said. I wouldn't be happy about it, but I'd have it although, like most doctors, I am a coward.

Whether I managed to reassure him or not, I do not know. It really was a simple operation, with a very low risk, but my registrar would have consented him earlier that morning and the registrars – especially the American ones – tend to go over the top with informed consent, and terrorize the poor patients with a long list of highly unlikely complications, including death. I mention the main risks as well but stress the fact that serious complications with simple disc prolapse surgery, such as nerve damage and paralysis, are really very rare.

I left the anaesthetic room to go to the meeting with the EWTD compliance manager.

'I'll come back and join you,' I said to my registrar over my shoulder as I left the theatre, though I thought that would scarcely be necessary as he had done such operations before on his own. I went back to the meeting room where my colleagues were waiting with the manager.

She was a large and officious young woman with hennaed hair in tight curls. She spoke imperiously.

'We need your agreement to the new rota,' she was saying.

'Well, what are the options?' one of my colleagues said.

'If they are to be compliant with the European Working Time Directive your registrars can no longer be resident on-call. The on-call room will be taken away. We have examined their diary cards – they are working far too much at the moment. They must have eight hours sleep every night, six of it guaranteed uninterrupted. This can only be achieved if they work in shifts like the SHOs.'

My colleagues stirred uncomfortably in their seats and grumbled.

'Shifts have been tried elsewhere and are universally unpopular,' one of them said. 'It destroys any continuity of care. The doctors will be changing over two or three times every day. The juniors on at night will rarely know any of the patients, nor will the patients know them. Everybody says it's dangerous. The shorter hours will also mean that they will have much less clinical experience and that's dangerous also. Even the President of the Royal College of Surgeons has come out against shifts.'

'We have to comply with the law,' she said.

'Is there any choice?' I asked. 'Why can't we derogate? Our juniors want to opt out of the EWTD and work longer

hours than forty-eight hours a week and can do this by der-
ogating. Everybody in the City opts out of the EWTD. My
medical colleagues in France and Germany say that they take
no notice of the EWTD. Ireland has derogated for doctors.'

'We have no choice,' she replied. 'Anyway, the deadline
for derogation was last week.'

'But we were only told last week about the possibility of
derogation!' I said.

'Well it's irrelevant anyway,' came the reply. 'The Trust
has decided nobody will derogate.'

'But that was never discussed with us. Does our opinion
about what is best for patients count for nothing?' I asked.

Her utter lack of interest in what I said was very obvious and she did not bother to reply. I started to deliver an
impassioned denunciation of the dangers of having trainee
surgeons working only forty-eight hours a week.

'You can send me an email setting out your views,' she
said, interrupting me, and the meeting came to an end.

I went round to the theatres where my registrar was starting the spinal case. He had done a fair number of these
cases on his own before, and although not the best of my
trainees in terms of operating ability, he was certainly one
of the most conscientious and kindest juniors I had had
for a long time. The nurses all adored him. It seemed safe
enough to let him start and probably do all the operation
himself. The patient's extreme anxiety had, however, made
me anxious in turn, so I changed and went into the theatre,
when usually I would have stayed outside in the red leather
sofa room, readily available but not overlooking everything
he did.

As it was a spinal procedure the patient, rendered anonymous by light blue sterile drapes, was lying anaesthetized
face down on the table, a small area of skin over the lower
spine exposed as a rectangle, coloured yellow by the iodine

antiseptic and brilliantly illuminated by the big, dish-shaped operating lights suspended on hinged arms from the ceiling. In the middle of this rectangle was a three inch incision through the skin and into the dark red spinal muscles, which were held open by steel retractors.

'Why such a large incision?' I asked irritably, still enraged by the manager and her complete indifference to what I had said. 'Haven't you seen how I do these? And why are you using the big bone rongeurs? That shouldn't be necessary at L5/S1.' I was annoyed but not alarmed – the operation had scarcely begun, the scan had showed a simple disc prolapse and he would not yet have reached the more difficult part of the operation, which is to expose the trapped nerve root within the spine.

I scrubbed up and came over to the operating table.

'I'll have a look,' I said. I picked up a pair of forceps and looked into the wound. A long shiny white thread, the thickness of a piece of string – four or five inches long – came up out of the wound in my forceps.

'Oh Jesus fucking Christ!' I burst out. 'You've severed the nerve root!' I threw the forceps onto the floor and flung myself away from the operating table to stand against the far wall of the theatre. I tried to calm myself down. I felt like bursting into tears. It is, in fact, highly unusual for gross technical mistakes like this to occur in surgery. Most mistakes during operations are subtle and complex and scarcely count as mistakes. Indeed, in thirty years of neurosurgery I'd never witnessed this particular disaster, although I have heard of it happening.

I forced myself to return to the operating table and looked into the bloody wound, cautiously exploring it, dreading what I might find. It became apparent that my registrar had completely misunderstood the anatomy and opened the spine at the outer rather than the inner edge of the spinal canal and hence had immediately encountered a nerve root, which, even more incomprehensibly, he had then severed. It was an utterly bizarre thing to have done, especially as he had seen dozens of these operations done before, and done many un-supervised on his own.

'I think you've cut straight through the nerve – a complete neurotmesis,' I said sadly to my dumb-struck assistant. 'He'll almost certainly be left with a permanently paralysed ankle and a life-long limp. That's not a minor disability – he'll never be able to run again, or to walk on uneven ground. So much for the mountain bike championships.'

We completed the rest of the operation in silence.

I redirected the opening into the spine and quickly re-moved the disc prolapse without any difficulty – the simple and quick operation I had more or less promised him as he lay looking so frightened in the anaesthetic room earlier that morning.

I went out of the theatre where Judith, my anaesthetist for many years, joined me in the corridor.

'Oh it's so terrible,' she said. 'And he's so young. What will you tell him?'

'The truth. It's just possible that the nerve is not com-pletely cut through and I suppose he might just recover, though if he does it will take months. To be honest, I doubt if he will, but I suppose there's some hope ...'

One of my consultant colleagues passed by and I told him what had happened.

'Bloody hell,' he said. 'That's bad luck. Do you think he'll sue?'

'I think it was reasonable enough for me to let my reg-istrar start – he's done these operations before. But I got it wrong. He was less experienced than I realized. It really was staggeringly incompetent ... but then I am responsible for his operating.'

'Well, it's the Trust that gets sued anyway – it doesn't really matter whose fault it was.'

'But I misjudged his abilities. I'm responsible. And the patient will blame me anyway. He'd put his trust in *me*, not in the bloody Trust. In fact, assuming he doesn't recover, I'll tell him to sue.'

My colleague looked surprised. Litigation is not something we are supposed to encourage.

'But my duty is to him, not to the Trust – isn't that what the GMC piously tells us doctors?' I said. 'If he's been left crippled and somebody's made a mistake – he ought to get some financial compensation, shouldn't he? The irony is that if we hadn't had to have that meeting with that fatuous manager I'd have been in theatre sooner and this disaster probably wouldn't have happened. I wish I could blame the manager,' I added. 'But I can't.'

I went off to write an operating note. It's quite easy to lie if things go wrong with an operation. It would be impossible for anybody to know after the operation in what way it had gone wrong. You can invent plausible excuses – besides, patients are always warned that nerve damage can happen with this operation, even though I have scarcely ever seen it happen. I know of at least one very famous neurosurgeon, now retired, who covered up an even more major mistake on a very eminent patient with a dishonest operating note. I wrote down, however, an exact and honest account of what had happened.

I left the theatre and thirty minutes later saw Judith leaving the recovery ward.

'Awake?' I asked.

'Yes. He's moving his legs ...' she said a little hopefully.

'It's the ankle that matters,' I replied gloomily, 'not the legs.'

I went round to see the patient. He was only just awake,

and was not going to remember anything I said so soon after the operation, so I said little to him and just sadly confirmed my worst fears: he had complete paralysis of lifting the left foot upwards – a foot drop as it is called in the trade – and, as I had told my junior, it is a very disabling condition.

I went to see him two hours later after he had returned to the ward and was fully awake. His wife was sitting anxiously beside him.

'The operation was not straightforward after all,' I said. 'One of the nerves for your left ankle was damaged and that's why you can't bend the foot up at the moment. It might get better – I really don't know. But if it does I'm afraid it will be a slow process that takes months.'

'But it should get better?' he asked anxiously.

I told him that I didn't know and could only promise to always tell him the truth. I felt quite sick.

He nodded in numb agreement, too shocked and confused to say anything. The anger and tears, I thought as I walked away, and dutifully squirted alcohol gel on my hands from a nearby bottle on the wall, will come later.

I went downstairs to my office and dealt with mountains of unimportant paperwork. There was a huge box of chocolates on my desk from a patient's wife. I took them through to Gail's office in the next room as she likes chocolates more than I do. Her office, unlike mine, has a window, and I noticed that it was pouring with rain in the hospital car park outside. The pleasant smell of rain on dry earth was filling her office.

'Have some chocolates,' I said.

I cycled home in a furious temper.

Why don't I just stop training juniors? I said to myself as I angrily turned the pedals. Why don't I just do all the operating myself? Why should I have to carry the burden of deciding whether they can operate or not when the fucking management and politicians dictate their training? I've got to see the patients every day on the ward anyway as the juniors are so inexperienced now – on the few occasions when they're actually in the hospital, that is. Yes, I shall no longer train anybody, I thought with a sudden sense of relief. It's not safe. There are so many consultants now that having to come in occasionally at night wouldn't be a great hardship ... The country's massively in debt financially, why not have a massive debt of medical experience as well? Let's have a whole new generation of ignorant doctors in the future. Fuck the future, let it look after itself, it's not my responsibility. Fuck the management, and fuck the government and fuck the pathetic politicians and their fiddled expenses and fuck the fucking civil servants in the fucking Department of Health. Fuck everybody.

15

MEDULLOBLASTOMA

n. a malignant brain tumour that occurs during childhood.

There was a child – Darren – who I had operated on many years ago for a malignant tumour called a medulloblastoma when he was twelve years old. The tumour had caused hydrocephalus and although I had removed the tumour completely the condition continued to be a problem and a few weeks after the operation I had carried out a 'shunt' operation, implanting a permanent drainage tube into his brain. My son William had undergone the same operation after his tumour had been removed for the same reason. William has been fine ever since but Darren's shunt had blocked on several occasions – a frequent problem with shunts – and he had required several further operations to revise the shunt. He was treated with radiotherapy and chemotherapy and as the years passed it appeared that he had been cured. Despite the problems with the shunt Darren had otherwise done very well and he went on to study accountancy at university.

He had been at university, away from home, when he started to develop severe headaches. He was brought to my hospital while I was on sick leave with a retinal detachment. A brain scan showed that the tumour had recurred. Although tumours such as Darren's can and do recur it is usually within the first few years after treatment. For the tumour to come back after eight years – as with Darren – is very unusual and nobody had expected it. Recurrence is inevitably fatal although further treatment can postpone death by a year or two with luck. The plan was that one of my colleagues would operate again in my absence but the evening before the operation Darren suffered a catastrophic haemorrhage into the tumour – an entirely unpredictable event that happens occasionally with malignant tumours. Even if he had been operated upon successfully before the haemorrhage it is unlikely that he would have had long to live. His mother had been with him when he had suffered the haemorrhage. He had been placed on a ventilator on the ITU but he was already brain dead and the ventilator was switched off a few days later.

I had got to know Darren and his mother well over the years and I had been very upset to hear of his death when I got back to work, though it was not the first time a patient of mine had died like this. As far as I could make out his treatment once he arrived in my department had been entirely appropriate but his mother was convinced he had died because of my colleague's delay in operating upon him. I received a letter from his mother requesting an appointment with me. I arranged to see her in my office rather than in one of the impersonal consulting rooms of the outpatient clinic. I brought her into the room and sat her down opposite me. She burst into tears and started to tell me the story of her son's death.

'He suddenly sat up in bed and clutched his head. My son cried out "Help me, help me, Mummy!"' she said, in torment as she told me. I remembered how once a patient of mine, dying from a tumour, had cried out for help to me and how awful and helpless I had felt. How much worse, I thought, how utterly unbearable it must be if it were one's own child crying for help, and if one could not help them.

'I *knew* that they should have operated but they just wouldn't listen to me,' she said.

She went over the sequence of events over and over again. After forty-five minutes I threw my hands up in the air and shouted in some desperation.

'But what do you want me to do? I wasn't there.'

'I know it wasn't your fault but I was hoping for some answers,' she replied.

I told her that as far as I could tell the haemorrhage could not have been predicted and it had been perfectly reasonable to plan on operating the next day. I said that the doctors and nurses who had been looking after Darren were terribly upset about what had happened.

'That's what they said on the ITU when they wanted to turn the ventilator off,' his mother said, her voice choking with anger. 'That keeping him on the ventilator was upsetting for the staff. But these people are paid, *they are paid*, to do their job!' She became so angry that she rushed out of the room.

I followed her out of the hospital into the afternoon sunlight to find her standing in the car park opposite the main entrance

'I'm sorry that I shouted,' I said. 'I find this all very difficult.'

'I thought you would be furious when you heard about his death,' she said to me in a disappointed voice. 'I know that it's difficult for you ...' – she waved her arm at the building behind us – 'You have a duty to the hospital.'

'I'm not trying to cover up for anybody,' I replied. 'I don't like this place and have no loyalty to it whatsoever.' As we talked we had started to walk back to the steel and glass front entrance to the hospital. The constant passage of people coming and going through the automatic doors made it feel like a railway station.

I took her back to my office, past the threatening notice at the entrance to the outpatient clinic over which I had once got into trouble for denouncing on the radio. 'This Trust' – states the notice – 'operates a policy of withholding treatment from violent and abusive patients ...' It was ironic, I thought, how the notice expressed the hospital management's distrust of patients, and it was a corresponding lack of trust in the hospital which was now tormenting Darren's mother. She collected her bag from my office and left without saying anything more.

I went back up to the wards. I met one of my registrars on the staircase.

'I've just seen Darren's mother,' I said to him. 'It was pretty grim.'

'There had been a lot of problems when the boy was dying on the ITU,' he replied. 'She wouldn't let us turn the ventilator off, even though he was brain dead. I had no problem with that, but some of the anaesthetic staff got pretty difficult over the weekend and some of the nurses were refusing to look after him since he was brain dead ...'

'Oh dear,' I said.

I remembered how angry I had been myself many years ago, at how my own son had almost died due to what I felt had been the carelessness of one of the doctors looking after him when he had been admitted to hospital with his brain tumour. I also remembered how, after I had become a neurosurgeon myself, I had operated on a young girl with a large brain tumour. The tumour was a mass of blood vessels, in the way that some brain tumours can be, and I had struggled desperately to stop the bleeding. The operation became a grim race between the blood pouring out of the child's head and my poor anaesthetist Judith pouring blood back in through the intravenous lines as I tried, and failed, to stop the bleeding.

The child, a very beautiful girl with long red hair, bled to death. She 'died on the table' – an exceptionally rare event in modern surgery. As I completed the procedure, stitching together the scalp of the now dead patient, there was utter silence in the operating theatre. The normal sounds of the place – the chatter of the staff, the hissing of the ventilator, the bleeping of the anaesthetic monitors – had suddenly stopped. All of us in the theatre avoided each other's eyes in the presence of death and in the face of such utter failure. And as I closed the dead child's head I had to think about what to tell the waiting family.

I had dragged myself up to the children's ward, where the mother was waiting to see me. She would not have been expecting to hear this catastrophic news. I had found it very difficult to talk, but managed to convey what had happened. I had no idea how she might react, but she reached out to me and held me in her arms and consoled me for my failure, even though it was she who had lost her daughter.

Doctors need to be held accountable, since power corrupts. There must be complaints procedures and litigation, commissions of enquiry, punishment and compensation. At the same time if you do not hide or deny any mistakes when things go wrong, and if your patients and their families know that you are distressed by whatever happened, you might, if you are lucky, receive the precious gift of forgiveness. As far as I know Darren's mother did not pursue her complaint but I fear that if she cannot find it in her heart to forgive the doctors who looked after her son in his final illness she will be haunted forever by his dying cry.

16

PITUITARY ADENOMA

n. a benign tumour of the pituitary gland.

By the time that I became a consultant in 1987 I was already an experienced surgeon. I was appointed to replace the senior surgeon at the hospital where I had been training and as the senior surgeon wound down his career he had delegated most of his operating to me. Once you become a consultant you are suddenly responsible for your patients in a way that you never were as a junior and trainee. You come to look back on your years of training as an almost carefree time. As a trainee the ultimate responsibility for any mistakes that you might make are ultimately borne by your consultant and not by yourself. As I get older I find the self-assurance of many of my trainees, for whose mistakes I am responsible, a little irritating but I was no different myself once. This all changes when you become a consultant.

My first few months in the role passed without incident. I was then referred a man with acromegaly. The disease is caused by a small tumour in the pituitary gland producing excess growth hormone. The person's face slowly changes – it becomes heavy and block-like, not unlike the cartoon figure Desperate Dan in the *Dandy* magazine, with a massive jaw and forehead. The feet enlarge and the hands become large and spade-like. The changes in this patient's case were not especially severe, and often the changes are so gradual over so many years that most patients and their families do not notice them. If one knew he had the condition one might notice that his jaw was a little heavy. The high levels of growth hormone ultimately damage the heart and it is for that reason, not the cosmetic changes, that we operate. The operation is done through the nostril, since the pituitary gland lies beneath the brain at the top of the nasal cavities, and is usually simple and straightforward. There are, however, two major arteries next to the pituitary gland that can, if the surgeon is exceptionally unlucky, be damaged during the operation.

His wife and three daughters had all come with him to my office when I first saw him. They were Italian and had become extremely emotional when I said that surgery would be required. They were obviously a close and loving family. Despite their anxieties about the operation, they expressed great confidence in me. He was a particularly nice person – I had been in to see him on the Sunday evening before the operation and we talked happily together for a while. It is a pleasant feeling when your patient obviously trusts you so completely. I operated the next day and the operation went well. He awoke perfectly. I went round to see him late that evening, and his wife and daughters were full of praise and thanks, which I happily acknowledged. The next day some of the symptoms of acromegaly – the feeling that his fingers were swollen – were already a little better and on Thursday morning I went to see him before he went home.

When I went to his bed and spoke to him he looked blankly back at me and said nothing. I then noticed that his right arm was lying useless beside him. One of the nurses hurried up to the bedside.

'We were trying to find you,' she said. 'We think he must have had a stroke just a few minutes ago.' My patient and

I looked uncomprehendingly at each other. I could scarcely believe, and he could not understand, what was happening. I felt a bitter wave of dread and disappointment break over me. Struggling against this I did my best to reassure him (though he would not have understood the words) that all would be well. But a brain scan later that morning confirmed a major stroke in his left cerebral hemisphere. This must have been caused in some unknowable way by the operation. He was by now aphasic – utterly without language. He did not seem too distressed by this, so presumably had little awareness of the problem and was living in some strange language-less world like a speechless animal.

Forgotten memories of other patients I had reduced to this grotesque state in the past suddenly came back to me. A man with an aneurysm in his brain, one of the first such operations I had carried out on my own as a senior registrar; another was an operation I had done on a man with a blood-vessel malformation in his brain. Unlike with this man, where the stroke occurred three days after the operation, with both these patients the operations had gone badly and they had suffered major strokes during the procedures. They had both looked at me afterwards with the same terrible dumb anger and fear, a look of utter horror – unable to talk, unable to understand speech – the look of the damned in some medieval depiction of hell. With the second patient, I remember the intense relief when I came to work next morning to find that he had suffered a cardiac arrest – as though the sheer trauma of what had happened to him had proved too much for his heart. The resuscitation team were working away at him – they were clearly not achieving anything, so I told them to stop and leave him in peace. I do not know what happened to the other man other than that he survived.

At least the Italian man seemed merely puzzled, and looked at me with a vague and empty expression. I had many long and emotional conversations with the family later that day. This involved floods of tears and much embracing. It is diffi- cult to explain, let alone to understand, what it must be like to have no language – to be unable either to understand what is said to one, or put one's thoughts into words. After major strokes people can die from brain swelling, but this patient remained unchanged for forty-eight hours, and the next evening I assured the family that he would not die, although I did not know if he would regain his speech, and rather doubted it. Nevertheless, two days later, at one o'clock in the morning he deteriorated.

My young and inexperienced registrar rang me.

'He's gone off and blown both his pupils!' he excitedly told me.

'Well, if both pupils have blown that means he's coned. He's going to die. There's nothing to be done,' I told him. Coning refers to the way in which the brain is squeezed like toothpaste out of the hole in the base of the skull when the pressure in the skull becomes very high. The extruded part of the brain is cone-shaped. It is a fatal process.

I went to bed, having growled to my registrar that I was not going to come in. But I couldn't get to sleep and instead drove in to the hospital, the streets deserted apart from a single fox confidently trotting across the road in front of the hospital, summer rain falling. The empty hospital corridors were ringing with the family's cries, including the three-year-old grand-daughter's. So I gathered them all together and sat in a chair facing them and explained things and told them how sorry I was. The patient's wife was on her knees in front of me, clasping her hands, begging me to save her husband. This went on for half an hour or so – it felt longer. They came to accept the inevitability of his death, and perhaps even that it was better for him than to live without language.

I remember another time I had had a patient die from a

stroke after an operation. The family had sat staring at me, glaring at me and saying nothing as I tried to explain and to apologize. It was quite clear they hated me and felt that I had killed their father.

But this family was extraordinarily kind and considerate. His daughters said that they did not blame me, and that their father had had great confidence in me. Eventually we parted – one of the daughters brought the three-year-old grand-daughter to me, who had now stopped crying. She looked up at me with two large and dark eyes above her tear-stained cheeks.

'Kiss the doctor good night, Maria, and say thank you.'

Maria laughed happily as we rubbed each other's cheeks.

'Goodnight, sweet dreams, Maria,' I dutifully said.

My registrar had been watching all this. He thanked me for sparing him the painful task of talking to the family.

'Terrible job, neurosurgery. Don't do it,' I said as I went past him on my way to the door.

I passed the patient's wife, standing beside the public phone in the corridor, as I walked to the front door.

'Remember my husband, please think of him sometimes,' she said, reaching a despairing hand out to me. 'Remember him in your prayers.'

'I remember all my patients who die after operations,' I said, adding to myself as I left, 'I wish I didn't.'

I was relieved that he had died – if he had survived he would have been left terribly disabled. He had died because of the operation but not as a result of any obvious mistake on my part. I do not know why the stroke had happened or what I could have done to avoid it. So, just for once, I felt, at least in theory, innocent. But when I arrived home I sat in my car outside my house, the rain falling in the dark, for a long time, before I could drag myself off to bed.

17

EMPYEMA

n. a condition characterized by an accumulation of pus in a body cavity.

It was a simple list: a craniotomy for a tumour, with a couple of routine spinal operations to follow. The first patient was a young man with a glioma on the right side of his brain that could not be removed completely. I had operated for the first time five years ago. He had remained perfectly well but his follow-up brain scans showed that the tumour was starting to grow back again and further surgery was now required, which would hopefully keep him alive for a few more years. He was unmarried and running his own business in IT. We got on well together whenever we met in the outpatient clinic and he had taken the news that he now needed further surgery with remarkable composure.

'We can hope that another op will buy you some extra years,' I told him. 'But I can't promise it … It might be much less. And the operation is not without some risks.'

'Of course you can't promise, Mr Marsh,' he replied.

I carried out his operation under local anaesthetic, so that I could check directly – simply by asking him – that I was not producing any paralysis down the left side of his body. When I tell a patient that I think I should do their operation under local anaesthetic they usually look a little shocked. In fact the brain cannot itself feel pain since pain is a phenomenon produced within the brain. If my patients' brains could feel me touching them they would need a second brain somewhere to register the sensation. Since the only parts of the head that feel pain are the skin and muscles and tissues outside the brain it is possible to carry out brain surgery under local anaesthetic with the patient wide awake. Besides, the brain does not come with dotted lines saying 'Cut here' or 'Don't cut there' and tumours of the brain usually look, more or less, like the brain itself, so it is easy to cause damage. If – as was the case here – the tumour was growing near the movement area in the right side of his brain that controlled the left side of his body, the only certain way I had of knowing if I was doing any damage while I operated was by having him awake. It is much easier to carry out brain surgery under local anaesthetic than you might think, provided that the patient knows what to expect, and trusts the surgical team – especially the anaesthetist who will look after the patient while the operation proceeds.

This man coped especially well and while I worked away he talked happily with my anaesthetist Judith – they remembered each other from the first operation and it was like listening to two old friends as they talked about holidays and families and recipes (he was a keen cook), while every few minutes Judith would ask him to move his left arm and leg and make sure that he could still move them as I worked on his brain with my sucker and diathermy.

So it was indeed a straightforward operation and after supervising my registrar with the two spinal cases I walked round to the ITU to see that he was fine, chatting to the nurse looking after him. I left the hospital to travel in to central London where I had a conference to attend.

*

I took my folding bicycle on the train to Waterloo. It was a singularly cold day with freezing rain and the city looked bleak and grey. I cycled to the legal chambers off Fleet Street where the conference was to be held. The case was over an operation I had carried out three years earlier. The patient had developed a catastrophic streptococcal infection afterwards, called a subdural empyema, which I had initially missed. I had never encountered a post-operative infection like this before and did not know any other neurosurgeons who had either. The operation had gone so well that I had found it impossible to believe it might all go wrong and I dismissed the early signs of the infection, signs which in retrospect were so painfully obvious. The patient had survived but because of my delay in diagnosing the infection she had been left almost completely paralysed and will remain so for the rest of her life. The thought of the conference had been preying on my mind for many weeks.

I presented myself to the receptionist in the grand and imposing marble lobby and was ushered into a waiting room. I was soon joined by a fellow neurosurgeon I know well who was advising my Defence Union over the case.

I told him about how I had come to make such a disastrous mistake.

Her husband had rung me up on my mobile phone on a Sunday morning when I was in the hospital dealing with an emergency. I didn't really take in what he said and misdiagnosed the infection as harmless inflammation. I should never have diagnosed that on the basis of a phone call but I was busy and distracted and I'd never had a serious complication with that particular operation before in twenty years.

'There but for the grace of God it could have been me,' my colleague said, trying to cheer me up. We were then joined by two solicitors from the Defence Union. They were very polite but quite without smiles. I thought that they looked tense and drawn but perhaps this was simply my imagination, produced by my awful feeling of guilt. I felt as though I was attending my own funeral.

We were taken downstairs to a basement room where a courteous QC – many years younger than me – was waiting for us. A large wall display extolled the virtues of his chambers in fine Roman capital lettering. I cannot remember what was claimed – I was too miserable to take much in.

Coffee was served and one of the solicitors unpacked box after box of documents onto the table.

'It's terrible how much trouble one phone call can cause,' I said sadly as I watched her, and she now smiled briefly at me.

'I need to start', the QC said very gently, 'to explain where we are coming from. I think this will be difficult to defend ...'

'I entirely agree,' I interrupted.

The meeting only lasted a couple of hours. It was painfully clear – as I had always known – that the case could not be defended.

At the end of the meeting the barrister asked my colleague to leave.

'Mr Marsh, perhaps you could stay behind,' he said.

I remembered once having to wait outside the office of my school headmaster fifty years ago, sick with anxiety, to be punished by the kind old man for some misdemeanour. I knew that the barrister was going to be professional and matter-of-fact but I nevertheless felt overcome with dread and shame.

After my colleague had left he turned to me. 'I'm afraid I don't really think we have a case here,' he said with an apologetic smile.

'I know,' I said. 'I've felt it was an indefensible mistake all along.'

'I'm afraid this might all drag on for a while,' one of the

solicitors added, sounding as I suspect I must sound when I break bad news to my patients.

'Oh that's all right,' I said, trying to sound brave and philosophical. 'I'm reconciled to this. It's neurosurgery. I'm just sorry to have wrecked the poor woman and to have cost you millions of pounds.'

'That's what we're here for,' she said. The three of them looked at me with kind, slightly questioning expressions. Perhaps they expected me to burst into tears. It felt strange to be an object of pity myself.

'Well, I'll leave you to discuss the awful financial consequences.' I said and picked up my satchel and folding bike.

'I'll see you to the door,' said the barrister and insisted on showing me the professional courtesy of accompanying me to the lift in the corridor outside. I did not feel that I deserved it.

We shook hands and he returned to discuss quantum, as lawyers call it – the cost of the settlement – with the two solicitors.

I found my colleague waiting for me in the lobby.

'It's the professional shame that hurts the most,' I said to him. I wheeled my bike as we walked along Fleet Street. 'Vanity really. As a neurosurgeon you have to come to terms with ruining people's lives and with making mistakes. But one still feels terrible about it and how much it will cost.'

The weather forecast had promised a dry morning and neither of us were dressed appropriately. Our professionals' pinstripe suits were getting soaked as we crossed Waterloo Bridge. As the rain streamed off my face my cheeks turned to ice.

'I know one has to accept these things,' I went on lamely, 'But nobody, nobody other than a neurosurgeon understands what it is like to have to drag yourself up to the ward and see, every day – sometimes for months on end – somebody one has destroyed and face the anxious and angry family at the bedside who have lost all confidence in you.'

'Some surgeons can't even face going on those ward rounds.'

'I told them to sue me. I told them that I had made a terrible mistake. Not exactly the done thing, is it? So I remained – crazily enough – quite good friends with them. At least I think so but I can't expect them to have a very high opinion of me, can I?'

'You can't stay pleased with yourself for long in neurosurgery,' my colleague said. 'There's always another disaster waiting round the corner.'

We walked into Waterloo Station, where the crowds were gathering to head south for the weekend, and shook hands and went our respective ways.

I had not dared to ask for how many millions of pounds the case would probably be settled. The final bill, I learned two years later, was for six million.

Back at the hospital that evening I went up to the ITU to see the young man with a recurrent tumour whom I had operated upon in the morning – it already felt like a lifetime ago. The operation had gone well enough but we both knew that I had not cured him and that the tumour would grow back again, sooner or later. He was sitting up in his bed, with a lopsided bandage around his head.

'He's fine,' said the nurse looking after him as she looked up from the lectern at the end of his bed where she was writing down the observations.

'Once again, Mr Marsh,' my patient said, looking at me intensely, 'My life in your hands. Really I can't thank you enough.' He wanted to say more but I put my finger to my lips.

'Ssshh,' I said, as I turned away to leave the ITU. 'I'll see you tomorrow.'

18

CARCINOMA

n. a cancer, esp. one arising in epithelial tissue.

I went to see my mother in hospital one Saturday. The cancer ward to which she had been admitted was on the tenth floor and her bed was beside a huge panoramic window. The view was of the Houses of Parliament and Westminster Bridge across the river, seen from above yet very close. The spring weather was exceptionally clear. The River Thames below us reflected the sunlight like polished steel and hurt my eyes. The city beyond was almost oppressive in its clarity – an unrelenting view of buildings, inhuman in scale and size – an inappropriate view, I thought, for somebody who was dying.

My mother said that the staff were very friendly but hopelessly overworked and disorganized compared to when she had been a patient in the same hospital many years earlier, gesturing to her bed which had been left unmade for two days as she said this. She hated complaining but admitted she had been kept starved for two consecutive days while waiting for an ultrasound scan – a scan which I knew was entirely unnecessary since she was already becoming jaundiced and obviously had metastases in her liver from the carcinoma of the breast for which she had been treated twenty years earlier. There was some light relief, she said, in using a commode while looking down on the nation's rulers across the river. She had grown up in Nazi Germany (from which she had escaped in 1939) and although a perfectly law-abiding citizen she was always sceptical of authority.

She was wasting away, as she herself commented. The bones of her face were becoming increasingly prominent and as she was stripped down to her underlying skeleton, I could recognize myself in her all the more clearly – people have always said that of her four children I am the closest to her in appearance. I could only hope that she might have a few good months left. We had an inconclusive discussion as to what she should do with what time she still had. My mother was one of the bravest and most philosophical people I had ever known but neither of us could bring ourselves to refer to death by name.

I was on call for the weekend and was telephoned endlessly by a new and inexperienced registrar about many difficult problems. These were not clinically difficult problems but problems caused by the chronic lack of beds.

On the following Monday there were various complaints from patients about my trying to discharge them from hospital too quickly. One of the complaining patients was a garrulous old man who was reluctant to go home with an in-dwelling urinary catheter after a simple spinal operation. I'd told him that he would be doing another patient a favour if he went home that day since we had no beds available for the patients who were due to be admitted for surgery next day. He was still on the ward three days later and the ward sister criticized me for having spoken to him in the way that I had (although I thought that I had been scrupulously polite). I had had to cancel an operation on a woman with severe trigeminal neuralgia because he wouldn't leave. The ward sister, nevertheless, told me that I had to apologize to him for having tried to get him to leave the hospital before he wished. So I went to say sorry to him (through silently gritted teeth). He accepted this happily.

'Yes, I understand, doctor,' he said. 'I used to work in the kitchen-fitting trade and sometimes couldn't complete a job on time. I also disliked disappointing people.'

I muttered something about brain surgery and building kitchen cupboards not being entirely equivalent and left his room – a balcony room, overlooking the hospital gardens and trees, with a distant view towards Epsom on the horizon. I was still working in the old hospital at the time – it was to be closed three years later. Perhaps if he had been in a more typical NHS bay, and not in a room on his own with a view of the hospital gardens and the many daffodils I had planted years before, he might have wanted to leave earlier.

I was away in Glasgow at a medical meeting two days later when the diagnosis of untreatable cancer was finally made and my mother was sent home to die. There was no question of chemotherapy in somebody her age with such advanced disease and she didn't want it either, which my father found hard to accept. I returned from Glasgow and went to my parents' house to find them sitting in the kitchen. My mother had become more jaundiced from liver failure since I had last seen her and looked worn and frail though fully herself.

'I don't want to leave you all,' she said sadly. 'But I don't think death is the end, you know.' My eighty-six-year-old father – already starting to suffer from the dementia from which he died eight years later – looked on, with a vague, lost expression as though he couldn't really take in what was happening: that his fifty-year-old son was crying over his wife of over sixty years, and that she was shortly to die.

Her condition deteriorated very quickly over the next few days, and she was dead within a fortnight; a short illness as the obituaries call it, though it felt quite long while it lasted. Until the very end she remained entirely lucid and completely herself, with her slightly ironical, modest sense of humour preserved until the last.

Each day she weakened, and soon she was established in a bed during the day in the music room downstairs. I would carry her up the staircase of my parents' house in my arms at the end of the day – by now she weighed next to nothing. But even this was quickly too much for her, and so after discussion with me and one of my sisters, who is a nurse, my mother remained in the bedroom she had shared with our father for the last forty years. This, she decided, would be where she would die. It was a beautiful room – a perfectly proportioned Georgian room with wood-panelling, painted a quiet, faded green, and an open fireplace and mantelpiece decorated with her collection of little pottery birds and eggs. The tall windows, with their rectangular panes of glass, looked out over the trees of Clapham Common, especially beautiful at that time of year. To the left one could see the church on the Common which she attended every Sunday and where her funeral service would be held.

Every morning and evening my sister and I would come in to care for her. At first I would help her to the bathroom where my sister would wash her but soon she was unable to walk even this short distance and instead I would lift her onto the commode we had borrowed from the local hospice. My sister was wonderful to watch, kindly and gently discussing and explaining everything as she carried out the simple, necessary nursing. We have both seen many people die, after all, and I had worked as a geriatric nurse many years ago too. It felt quite easy and natural for us both, I think, despite our intense emotions. It's not that we felt anxious – the three of us knew she was dying – I suppose what we felt was simply intense love, a love quite without ulterior motive, quite without the vanity and self-interest of which love is so often the expression.

'It's a quite extraordinary feeling to be surrounded by so much love,' she said two days before she died. 'I count my blessings.'

She was right to do this, of course. I doubt if any of us will enjoy – if that's the word to use – such a perfect death when our own time comes. To die in her own home, after a long life, quite quickly, looked after by her own children, surrounded by her family, entirely free of pain. A few days before she died, almost by chance, the family – children, grandchildren and even great-grandchildren, and two of her oldest friends – found themselves all gathered in the family home. We staged what amounted to an impromptu wake, before her death, much to my mother's delight. While she lay dying upstairs we sat down round the dining room table and remembered her life, and drank to her memory even though she was not yet dead, and ate supper cooked by my wife-to-be Kate. I had only met Kate – to my mother's joy, after the trauma of the end of my first marriage – a few months earlier. Kate had been slightly surprised to find herself cooking supper for seventeen people when earlier in the day I had hesitantly asked her if she might cook supper for five.

Each day I thought might be the last but each morning when I returned she would say to me, 'I'm still here.'

Once when I told her, as I kissed her goodnight, that I would see her in the morning, she replied with a smile,

'Dead or alive.'

My family were playing out an age-old scene that I suppose is rarely seen now in the modern world, where we die in impersonal hospitals or hospices, cared for by caring professionals, whose caring expressions (just like mine at work) will disappear off their faces as soon as they turn away, like the smiles of hotel receptionists.

Dying is rarely easy, whatever we might wish to think.

Our bodies will not let us off the hook of life without a struggle. You don't just speak a few meaningful last words to your tearful family and then breathe your last. If you don't die violently, choking or coughing, or in a coma, you must gradually be worn away, the flesh shrivelling off your bones, your skin and eyes turning deep yellow if your liver is failing, your voice weakening, until, near the end, you haven't even the strength to open your eyes, and you lie motionless on your death bed, the only movement your gasping breath. Gradually you become unrecognizable – at least you lose all the details that made your face characteristically your own, and the contours of your face are worn away down to the anonymous outlines of your underlying skull. You now look like the many old people, with drawn and dehydrated faces, identical in their hospital gowns, to whose bedside I would be summoned in the early hours when I worked as a junior doctor, down the long and empty hospital corridors, to certify death. Your face becomes that of Everyman, close to death, a face we all know, if only from the funeral art of Christian churches.

By the time she died my mother was no longer recognizable. I last saw her on the morning of the day she died before I set off to work. I had spent the night in my parents' house, sleeping on the floor of my father's study, near my parents' bedroom. I could hear her rasping breath through the open doors between the study where I was lying and her bedroom. When I went to see her at four in the morning she shook her head when I asked her if she wanted some water and morphine, even though one would have thought from her appearance that she was already dead, had it not been for her laboured, occasional breathing. Before I finally left I said to her, to her death mask, as I held her hand, 'You're still here.' Almost imperceptibly she slowly nodded her head. I cannot remember my last sight of her when I went to work in the morning – it no long mattered. I had said goodbye to her many times already.

My sister rang me shortly after midday, as I sat at some dull medical meeting, to say she had died a few minutes earlier. Her breathing, she told me, had become shallower and shallower until eventually my family, who were gathered round her bed realized, with slight surprise, that she had died.

I felt no need to pay her body my final respects – as far as I was concerned her body had become a meaningless shell. I say 'body' – I could just as well talk of her brain. As I had sat by her bedside I had often thought of that – of how the millions upon millions of nerve cells, and their near-infinite connections that formed her brain, her very self, were struggling and fading. I remember her on that last morning, just before I went to work – her face sunken and wasted, unable to move, unable to talk, unable to open her eyes – yet when I asked her if she wanted any water to drink she shook her head. Within this dying, ruined body, invaded by cancer cells, 'she' was still there, even though she was now refusing even water, and clearly anxious not to prolong her dying any longer. And now all those brain cells are dead – and my mother – who in a sense was the complex electrochemical interaction of all these millions of neurons – is no more. In neuroscience it is called 'the binding problem' – the extraordinary fact, which nobody can even begin to explain, that mere brute matter can give rise to consciousness and sensation. I had such a strong sensation, as she lay dying, that some deeper, 'real' person was still there behind the death mask.

What makes for a good death? Absence of pain, of course, but there are many aspects to dying and pain is only one part of it. Like most doctors I suppose I have seen death in all its many forms and my mother was indeed lucky to die in the way that she did. If I ever think about my own death – which, like most people, I try to avoid – I hope for a quick end, with a heart attack or stroke, preferably while asleep. But I realize that I may not be so fortunate. I may very well have to go through a time when I am still alive but have no future to hope for and only a past to look back on. My mother was lucky to believe in some kind of life beyond death but I do not have this faith. The only consolation I will have, if I do not achieve instant extinction, will be my own last judgement on my life as I look back on it. I must hope that I live my life now in such a way that, like my mother, I will be able to die without regret. As my mother lay on her death-bed, drifting in and out of consciousness, sometimes lapsing into her German mother-tongue she said:

'It's been a wonderful life. We have said everything there is to say.'

19

AKINETIC MUTISM

n. a syndrome characterized by the inability to speak, loss of voluntary movement and apparent loss of emotional feeling.

Neuroscience tells us that it is highly improbable that we have souls, as everything we think and feel is no more or no less than the electrochemical chatter of our nerve cells. Our sense of self, our feelings and our thoughts, our love for others, our hopes and ambitions, our hates and fears all die when our brains die. Many people deeply resent this view of things, which not only deprives us of life after death but also seems to downgrade thought to mere electrochemistry and reduces us to mere automata, to machines. Such people are profoundly mistaken, since what it really does is upgrade matter into something infinitely mysterious that we do not understand. There are one hundred billion nerve cells in our brains. Does each one have a fragment of consciousness within it? How many nerve cells do we require to be conscious or to feel pain? Or does consciousness and thought reside in the electrochemical impulses that join these billions of cells together? Is a snail aware? Does it feel pain when you crush it underfoot? Nobody knows.

An eminent and eccentric neurologist who had sent me many patients over the years asked me to examine a woman

I had operated on a year earlier who was in a persistent vegetative state. I had operated for a ruptured arterio-venous malformation after she had suffered a life-threatening haemorrhage and I had operated as an emergency. It had been a difficult operation and although it had saved her life it could not undo the damage done to her brain by the haemorrhage. She had been in a coma before the operation and remained in a coma for many weeks afterwards. She had been transferred back to her local hospital some weeks after the operation where she had been under the care of the neurologist who now wanted me to see her in the long-term nursing home in which she had ended up. Before she was transferred to the nursing home, I had carried out a shunt operation for hydrocephalus which had developed as a late after-effect of the original bleed.

Although the shunt operation had been a relatively minor one – one I would usually delegate to my juniors – I remembered it well because I had carried it out at the local hospital and not in my own neurosurgical centre. I scarcely ever operate away from my own theatres, except when I am working abroad. I had gone to the local district hospital where she was a patient with a tray of instruments and one of my registrars. I had gone vainly thinking that the visit of a senior neurosurgeon to the hospital – since brain surgery was not normally performed there – would be an event of some importance and be of some interest, but apart from the desperate family everybody else in the hospital seemed scarcely to notice my arrival. The local neurologist, who was away at the time of my visit, had told the family that the operation might relieve her persistent vegetative state. I was less optimistic, and said so, but there was little to be lost by trying and so after discussing this with them I went down to the operating theatres where, I was told, the staff were ready for me to operate.

The nurses and the anaesthetists greeted me with total indifference, which I found quite disconcerting. I had to wait two hours before the patient was brought down for the operation, and when she eventually came into the theatre, the staff all worked in sluggish and sullen silence. The contrast to my own friendly and energetic neurosurgical theatres was remarkable. I had no way of knowing whether they felt that I was wasting their time by operating on a human vegetable or whether this was just their normal way of behaving. So I operated, reported back to the family afterwards, and drove back to London.

As the months passed after this second operation, it became clear that the shunt had made no difference to her condition, and her neurologist wanted me to examine the patient and see if the shunt was working or if it had blocked. It seemed a little cruel and unnecessary to drag her all the way to my hospital in an ambulance just for my opinion so I had agreed – a little reluctantly, as I knew that I could not help – to visit her in the nursing home that now cared for her.

Patients in persistent vegetative state – or PVS as it is called for short – seem to be awake because their eyes are open, yet they show no awareness or responsiveness to the outside world. They are conscious, some would say, but there is no content to their consciousness. They have become an empty shell, there is nobody at home. Yet recent research with functional brain scans shows this is not always the case. Some of these patients, despite being mute and unresponsive, seem to have some kind of activity going on in their brains, and some kind of awareness of the outside world. It is not, however, at all clear what it means. Are they in some kind of perpetual dream state? Are they in heaven, or in hell? Or just dimly aware, with only a fragment of consciousness of which they themselves are scarcely aware?

There have been several high-profile court cases in recent

years as to whether treatment that keeps these people alive – since they cannot eat or drink – should be withdrawn or not, whether they should be left to die or not. In several cases the judges decided that it was reasonable to withdraw treatment and let the vegetative patients die. This does not happen quickly – instead the law, solemn and absurd, insists that the patients are slowly starved and dehydrated to death, a process that will take several days.

I finished my outpatient clinic at eight and drove out of London in the early autumnal evening. It was quite late by the time I reached the neurologist's home. He drove me in his own car to the nursing home a few miles away. This was a pleasant country house, surrounded by tall and ancient trees. It was night by now, and once we had parked the car I could see the friendly lights of the nursing home through the dark branches of the trees as we walked across a derelict tennis court covered in dry, fallen leaves. The home was run by Catholic nuns and devoted to people with catastrophic brain damage. Inside all was clean and tidy, and the staff were clearly very caring and friendly. The contrast with the hospital where I had carried out the shunt operation a year earlier could not have been greater. The devout Catholic staff did not accept the grave lesson of neuroscience – that everything we are depends upon the physical integrity of our brains. Instead, their ancient faith in an immaterial human soul meant that they could create a kind and caring home for these vegetative patients and their families.

The sister took me up a grand staircase to see my patient. I wondered who had lived in the house originally – an Edwardian capitalist perhaps or lesser aristocrat, with a small army of servants. I wondered what he would have thought of the use to which his imposing home had now been put. On the first floor there was a wide, carpeted corridor which we walked down, passing many patients in their rooms on either side. The doors were all open and through the doorways I could see the motionless forms of the patients in their beds. Beside each door was an enamelled plaque with the patient's name; because they are there for so many years, until they die, they can have proper plaques rather than just the paper labels you find in an ordinary hospital. To my dismay I recognized at least five of the names as former patients of mine.

One of the senior neurosurgeons who trained me, and a man I revere, told me a story once of the famous, knighted surgeon with whom he, in his turn, had served as an apprentice.

'He used to remove acoustic tumours with a periosteal elevator, an instrument normally used for opening the skull,' he told me. 'An operation that would take most surgeons many hours took him thirty or forty minutes. Inevitably this would sometimes lead to disaster. I remember one woman with a large acoustic – he caught the vertebral artery with the elevator and there was torrential haemorrhage. The woman was obviously done for. I had to close up and that was that. Nevertheless, I always had to ring him up every evening at seven o'clock on the dot to let him know how all the patients were doing. So I went through the list of all the inpatients. At the end I mentioned the woman with an acoustic. Mrs B she was called, I can still remember the name. Mrs B is slipping away, I said, or words to that effect. "Mrs B?" he said. "Who's that?" He had forgotten her already. I wish I had a memory like that,' my boss said wistfully. 'Great surgeons,' he then added, 'tend to have bad memories.'

I hope I am a good surgeon but I am certainly not a great surgeon. It's not the successes I remember, or so I like to think, but the failures. But here in the nursing home were several patients I had already forgotten. Some of them were people I had simply been unable to help, but there was at least one man who, as my juniors put it in their naive and tactless way, I had wrecked.

I had ill-advisedly operated on him many years earlier for a large tumour in a spirit of youthful enthusiasm. The operation had gone on for eighteen hours and I had inadvertently torn the basilar artery at two in the morning – this is the artery that supplies the brainstem, and he never woke up again. I saw his grey curled-up body in its bed. I would never have recognized him were it not for the enamelled plaque with his name by the door.

The patient I had come to see lay mute and immobile, limbs rigid, her eyes open in an expressionless face. She had been a journalist for a local newspaper, full of life and energy, but had then suffered the haemorrhage which caused the damage which my operation could not undo. There were happy, smiling photographs of her before this terrible event on the walls around the room. She made occasional mewling sounds. It took me only a few minutes to test the shunt by putting a needle into it through the skin of her scalp and establish that it was working. There was nothing I could do to help.

She communicated, apparently, via a Morse-code buzzer, as it seemed that she could move one finger. A nurse was sitting beside her and patiently listened to the bleeping sounds, concentrating very hard with a slight frown. She interpreted them for me. The patient asked me, the nurse explained, about the shunt and then she thanked me and wished me goodnight.

Her mother was there and came out of the room with me, accosting me a little desperately in the wide corridor outside. We spoke for a while. She talked about letters that her daughter had been sending – transcribed by one of the nurses from her Morse-code bleeping. She expressed some doubts as to whether her daughter had really said the things transcribed by the nurses.

There is no way of knowing, of course. The woman's

mother lives in a nightmare, a labyrinth of uncertainty and of hopeless love, her daughter both alive and dead. Behind her daughter's rigid, expressionless face, is she in fact awake? Is she aware, in some way, of what is going on outside her paralysed body? Are the nurses inventing her letters – wittingly or unwittingly? Does their faith deceive them? Can we ever know?

20

HUBRIS

n. arrogant pride or presumption; (in Greek tragedy) excessive pride towards or defiance of the gods, leading to nemesis.

I went to Marks & Spencer in Wimbledon in the morning and bought a boxful of fruit and chocolates for the theatre staff. I had gone through my CD collection and picked out enough disks to last at least all day and much of the night, as the operation was going to be a long one. I had only been a consultant for four years but I already had a very large practice, larger than any other neurosurgeon's I knew. The patient was a schoolteacher in his late fifties, tall and bespectacled, who walked with a stick and was a little stooped. He had been seen by a local neurologist who had arranged a brain scan and as a result he had been sent to see me. It was in the days of the old hospital and I saw him in my office, with its row of windows looking out onto a little copse of birch trees. One of the local foxes would sometimes look in at me with a thoughtful expression as it passed by. I sat the patient down in the chair by my desk with his wife and son next to him and took the films of his brain scans – which he had brought with him – over to the viewing box on the wall. Computers were still a long way away.

I already knew what the scans would show but I was

still startled by the sheer size of the tumour growing from the base of his skull. All of the brainstem and the cranial nerves – the nerves for hearing, movement, sensation for the face, and for swallowing and talking – were stretched over its sinister hump-backed mass. It was an exceptionally large petro-clival meningioma. I had only seen tumours this size before in the textbooks. In later years I was to see many such in Ukraine when patients with terrible tumours from all over the country would come to seek my opinion. I was not sure whether to feel excited or alarmed.

I went back to my desk and sat down next to him.

'What have you been told about this?' I asked.

'The neurologist said it was benign,' he replied 'And that it was up to you as to whether it should be removed or not.'

'Well, it's certainly benign but it's also very large,' I said. 'But they grow very slowly so we know it's been there for many years already. What led you to having the scan in the first place?'

He told me how he had noticed that his walking had slowly been becoming a little unsteady in recent years and that he was also starting to lose hearing in his left ear.

'But what will happen if it stays there?' his son asked.

I replied cautiously, telling them that it would go on grow-ing slowly and that he would slowly deteriorate.

'I've already decided to take early retirement on medical grounds,' he said.

I explained that surgery was not without risks.

'What sort of risks?' asked the son.

I told them that the risks were very serious. There were so many brain structures involved with the tumour that the dangers of surgery ranged from deafness or facial paralysis to death or a major stroke. I described what surgery would involve.

The three of them sat in silence for a while.

'I've been in contact with Professor B in America,' the
son said. 'He said it should be operated upon and he said he
could do it.'

I was not sure what to say. I was only at the beginning
of my consultant career and knew that other surgeons were
more experienced than I was. In those years I was in awe
of the big names of international neurosurgery, who you
could hear giving keynote lectures at conferences where they
showed cases like the man in front of me, and the amazing
results they achieved, quite beyond anything that I had yet
done.

'But it would cost over 100,000 dollars,' the patient's wife
added. 'And we can't afford that.'

The son looked a little embarrassed.

'We're told that Professor M is the best neurosurgeon
in the country,' he said 'And we're going to see him for a
second opinion.'

I felt humiliated but knew that any operation was going to
be exceptionally difficult.

'That's a good idea,' I said. 'I'd be very interested to hear
what he thinks.' They left the room and I continued with my
outpatient clinic.

'I've got Professor M on the line for you,' said Gail two
weeks later, looking into my office.

I picked up my phone to hear the professor's booming,
confident voice. I had known him briefly when I was a
trainee and he was certainly a superb surgeon whom all the
trainees hoped to emulate. Self-doubt had never seemed to
be one of his weaknesses. I had heard that he would soon be
retiring.

'Ah, Henry!' he said 'This chap with the petro-clival.
Needs to come out. He's starting to have some difficulties
with swallowing so it's only a matter of time before he gets

aspiration pneumonia and that will be the end of him. It's a young man's operation. I've told them you should do it.'

'Thank you very much, Prof,' I replied, a little surprised but delighted to have been given what felt like a papal dispensation.

So I made arrangements for the operation, which I expected to be a long one. This was many years ago, when hospitals were different places, and all I had to do was ask the theatre staff and anaesthetists to stay on longer than usual. There were no managers whose permission had to be sought. The operation started in an almost festive spirit. This was man-sized brain surgery – a 'real Big Hit' as the American registrar assisting me put it.

As we opened the man's head, we talked about the big names of neurosurgery in America.

'Prof B's a really fantastic surgeon, amazing technician,' my registrar said, 'but do you know what he was called by his residents before he moved to his present job? They called him "the Butcher" because he trashed so many patients as he perfected his technique with these really difficult cases. And he still gets some terrible complications. Doesn't seem to trouble him much though.'

It's one of the painful truths about neurosurgery that you only get good at doing the really difficult cases if you get lots of practice, but that means making lots of mistakes at first and leaving a trail of injured patients behind you. I suspect that you've got to be a bit of a psychopath to carry on, or at least have a pretty thick skin. If you're a nice doctor you'll probably give up, let Nature takes its course and stick to the simpler cases. My old boss, who was really nice – the one who operated on my son – used to say 'If the patient's going to be damaged I'd rather let God do the damage than do it myself'.

'In the US,' my registrar continued, 'we're a bit more

can-do, but we have a commercial health-care system and nobody can afford to admit to making mistakes.'

The first few hours of the operation went perfectly. We slowly removed more and more of the tumour, and by mid-night, after fifteen hours of operating, it looked as though most of it was out and the cranial nerves were not damaged. I started to feel that I was joining the ranks of the really big neurosurgeons. I would stop every hour or two, and join the nurses in the staff room and have something to eat and drink from the box I had brought and smoke a cigarette – I stopped smoking some years later. It was all very convivial. Music played continuously while we operated – I had brought all sorts of CDs in that morning ranging from Bach to Abba to African music. In the old hospital I always listened to music when operating and although my colleagues found some of my choices a little strange they seemed to like it, especially what we called 'closing music' which meant playing Chuck Berry or B.B. King or other fast rock or blues music when stitching up a patient's head.

I should have stopped at that point, and left the last piece of tumour behind, but I wanted to be able to say that I had removed all of the tumour. The post-operative scans shown by the big international names when they gave their keynote lectures never showed residual tumour so surely this was the right thing to do, even if it involved some risk.

As I started to remove the last part of the tumour I tore a small perforating branch off the basilar artery, a vessel the width of a thick pin. A narrow jet of bright red arterial blood started to pump upwards. I knew at once that this was a catastrophe. The blood loss was trivial, and easy enough to stop, but the damage to the brainstem was terrible. The basilar artery is the artery that keeps the brainstem alive and it is the brainstem that keeps the rest of the brain awake. As a result the patient never woke up and that was why, seven years later, I saw him curled into a sad ball, on a bed in the nursing home.

I will not describe the pain of seeing his unconscious form on the ITU for many weeks after the operation. To be honest I cannot remember it well now, the memory has been overlain by other, more recent tragedies, but I do remember many anguished conversations with the family as we all hoped against hope that he would wake up again one day.

It is an experience unique to neurosurgeons, and one with which all neurosurgeons are familiar. With other surgical specialties, on the whole, the patients either die or recover, and do not linger on the ward for months. It is not something we discuss among ourselves, other than perhaps to sigh and nod your head when you hear of such a case, but at least you know that somebody understands what you feel. A few seem to be able to shrug it off, but they are a minority. Perhaps they are the ones who will become great neurosurgeons.

Eventually the poor man was sent back to his local hospital, still in a coma but no longer on a ventilator, and at some point he had been sent on to a nursing home where he had remained ever since. This was the man I had seen and scarcely recognized on my visit to see the girl with akinetic mutism.

For the next few years whenever I saw similar cases – which was only on a few occasions – I deemed the tumours inoperable and left the unfortunate patients to go elsewhere or to have radiation treatment, which is not very effective for very large tumours of this kind. These were also the years when my marriage fell apart and the old hospital was closed. I am not sure whether I realized it then, but this was the time when I became a little sadder but, I would like to think, much wiser.

Nevertheless, I gradually regained my courage and used what I had learned from the tragic consequences of my hubris to achieve much better results with tumours of this kind. I would, if necessary, operate in stages over several weeks, I would operate with a colleague, taking the operating in turns with an hour on and an hour off, like drivers in a military convoy. I would not try to remove all of the tumour if it looked as though it would be particularly difficult. I would rarely let an operation take longer than seven or eight hours.

The problem remains, however, that such tumours are very rare. In Britain, with a culture which believes in the virtues of amateurism, and where most neurosurgeons are very reluctant to refer difficult cases on to a more experienced colleague, no individual surgeon will ever gain as much experience as some of our colleagues do in the US. In America there are far more patients, and therefore more patients with such tumours. The patients are less deferential and trusting than they are in Britain. They are more like consumers than petitioners, so they are more likely to make sure that they are treated by an experienced surgeon.

After twenty-five years I would like to think that I have become relatively expert – but it has been a very long, slow advance with many problems along the way, though none as awful as that first operation. A few years ago I operated on the sister of a famous rock musician with a very similar tumour and, after a difficult time for the first few weeks after the operation, she made a perfect recovery. Her brother gave me a large sum of money from the charitable fund he runs which has helped fund my work in Ukraine and elsewhere ever since, so perhaps I can say that some good came out of that wretched operation many years ago.

There were two other lessons that I learned that day. One was not to do an operation that a more experienced surgeon than me did not want to do; the other was to treat some of the keynote lectures at conferences with a degree of scepticism. And I can no longer bear to listen to music when operating.

21

PHOTOPSIA

n. the sensation of flashes of light caused by mechanical stimulation of the retina of the eye.

Illness is something that happens only to patients. This is an important lesson you learn early on as a medical student. You are suddenly exposed to a terrifying new world of illness and death, and you learn how terrible illnesses often start with quite trivial symptoms – blood on the toothbrush might mean leukaemia, a small lump in the neck might mean cancer, a previously unnoticed mole might be malignant melanoma. Most medical students go through a brief period when they develop all manner of imaginary illnesses – I myself had leukaemia for at least four days – until they learn, as a matter of self-preservation, that illnesses happen to patients, not to doctors. This necessary detachment from patients becomes all the greater when you start working as a junior doctor and you have to do frightening and unpleasant things to patients. It starts with simple blood-taking and inserting drips, and progresses over time – if you train as a surgeon – to ever more radical procedures, cutting and slicing into people's bodies. It would be impossible to do the work if you felt the patients' fear and suffering yourself. Besides, the increasing responsibility that comes as you climb the career ladder brings greater anxiety that you will make a mistake and that patients will suffer. Patients become objects of fear as well as of sympathy. It is much easier to feel compassion for other people if you are not responsible for what happens to them.

So when doctors fall ill themselves they tend to dismiss their initial symptoms and find it hard to escape the doctor–patient relationship, to become mere patients themselves. It is said that they are often very slow to diagnose their own illnesses. I took little notice of the flashing light in my eye. It had started when I returned to work in September after a late summer holiday. I noticed that every time I walked down the brightly lit, factory-like corridors of the hospital an odd little flashing light would momentarily appear in my left eye. It was hard to pin down and after a fortnight it disappeared. A few weeks later, however, I noticed that there seemed to be a flashing arc, just beyond my line of vision in my left eye, which would come and go for no apparent reason. I became a little pre-occupied by this but since the symptoms were almost subliminal I dismissed them, although I could not help but think of the patients I see whose brain tumours can sometimes first declare themselves with rather similar subtle visual symptoms. I attributed them instead to anxiety about the meeting to which I had been summoned with the hospital chief executive, probably to be told off for causing trouble again.

One evening, while driving my car, there was a sudden shower of flashing lights in my left eye, as swift as shooting stars. When I got home I found that my eye seemed to have filled with a swirling black cloud of Indian ink. It was rather alarming but quite painless. I had paid little attention to ophthalmology as a student and didn't have a clue as to what was happening but a few minutes on the internet revealed that I had suffered a vitreous detachment. The vitreous – the transparent jelly that fills the eye behind the lens – had broken away from the wall of the eyeball. Since I am very short-sighted I learned that I was at risk of the vitreous detachment progressing to a retinal detachment which might result in my losing vision in that eye.

A major advantage of being a doctor is that you can get immediate medical help from your friends without the misery that our patients face of queuing in the local Accident and Emergency department, in a GP's surgery or, worse still, trying to get hold of a GP out of hours. I rang up an ophthalmic colleague. He arranged to see me early next morning, a Sunday. So next day I drove to the hospital where we both work, the roads empty, the vision in my left eye intermittently blurred by the floating cloud of black blood. He examined my eye and told me that I had the beginnings of a detached retina. It was in the days when I still had a large private practice and could afford private medical insurance so it was arranged for me to see a specialist vitreo-retinal surgeon in one of the central London private hospitals the following day.

I knew by now that retinal detachment can occur quite suddenly – the retina can simply peel off the eyeball, like old wallpaper off a damp wall – and I lay that night in my dark bedroom, my wife Kate beside me as anxious as I was, opening and closing my eye, checking if I could see, wondering if the eye might go blind, watching the dim shape of the blood-cloud performing its dance across the night sky seen through the windows. It turned and twisted slowly, quite elegantly, a little like a computer screen saver. To my surprise I eventually got to sleep and could see well enough in the morning to go to work – the appointment with the vitreo-retinal surgeon was for the afternoon.

Surgeons can fall ill just like anybody else but it can be difficult to judge whether one is well enough to operate. You cannot cry off operating just because you are feeling a little out of sorts but nor would anybody want to be operated upon by a sick surgeon. I learned a long time ago that I can operate perfectly well despite being tired, as when I am operating I am in an intense state of arousal. Sleep deprivation research has shown that people make mistakes if moderately deprived of sleep when they are carrying out boring, monotonous tasks. Surgery – however trivial the operation – is never boring or monotonous. I carried out one operation – ironically enough under local anaesthetic on the visual area of a man's brain – and quite forgot my own anxieties until, as I started to put his skull back together again, I remembered that I was soon to be a patient myself.

Suddenly fearful, I left the hospital in a hurry, and ordered a minicab to take me to the Harley Street Clinic in central London.

The retinal surgeon was a little younger than me but I recognized myself in his surgical manner – affable and business-like, with that wary sympathy all doctors develop, anxious to help but worried that patients will make difficult emotional demands of us. I knew that he would dislike having to treat a fellow surgeon – it is both a compliment and a curse when your colleagues ask you to treat them. All surgeons feel anxious when treating colleagues. It is not a rational anxiety – their colleagues are much less likely to complain than other patients if things go badly, as they know all too well that doctors are fallible human beings and not entirely in control of what is going to happen. The surgeon treating a fellow surgeon feels anxious because the usual rules of detachment have broken down and he feels painfully exposed. He knows that his patient knows that he is fallible.

He examined my retina again. The light was especially bright and I flinched a little.

'There's fluid starting to build up under the retina,' he said. 'I'll operate tomorrow morning.'

I walked out of the building twenty minutes later in a state of panic. Rather than take the tube or a taxi home I walked the six miles back to my house rehearsing all the terrible things that might happen to me — starting with having to abandon my career (I did, in fact, know two surgeons who had had do this because of retinal detachments) and going on to complete blindness, which was possible, since I had been told I had early changes predictive of detachment in my other eye as well. I cannot remember how my thoughts ran as I walked but, to my surprise, by the time I got home I was strangely reconciled to the problem. I would accept whatever happened but hoped for the best. I had forgotten that I had turned off my mobile phone when in the clinic and I shame-facedly found a panic-stricken Kate waiting at home, fearing the worst, unable to contact me.

At the hospital the next morning a smart receptionist was expecting me. The paperwork was quickly dealt with and I was taken to my room. The porters and attendants wore black waistcoats like page boys, the corridors and rooms were all carpeted and quiet with muted lighting. The contrast with the large public hospital where I work couldn't have been greater. The surgeon re-examined my left eye and told me that I needed an operation called a gas-bubble vitrectomy in which several large needles are inserted into the eyeball, the vitreous jelly is sucked out and the retina plastered back into place with an ice-cold cryo-probe. The eyeball is then filled with nitrous oxide gas to keep the retina in place for the next few weeks.

'You can have the operation under local or general anaes-thetic,' the surgeon told me, in a slightly hesitant voice. It was clear that he did not find the idea of operating on me under local anaesthetic appealing and neither did I, though I felt a coward when I thought of how I subject many of my patients to brain surgery under local anaesthetic.

'General anaesthetic, please,' I said to his evident relief and then his anaesthetist, who must have been waiting outside with his ear to the door, bounced into the room like a jack-in-the-box and quickly checked my fitness for an anaesthetic. Half an hour later, dressed in one of those absurd gowns that for some obscure reason fasten at the back, rather than at the front, usually leaving one's buttocks exposed, with paper knickers, white anti-embolism stockings and a pair of well-used slippers, I was being escorted to the operating theatres by one of the nurses. As I walked into the anaesthetic room I almost burst out laughing. I must have walked into operating theatres thousands of times, the all-important surgeon, in charge of his little kingdom, and here I now was as the patient, dressed in gown and paper knickers.

I had always dreaded becoming a patient yet when, at the age of fifty-six, I eventually did I found it remarkably easy. This was, quite simply, because I realized how lucky I was compared to my own patients – what could be worse than having a brain tumour? What right did I have to complain when others must suffer so much more? Perhaps it was also because I was using my private health insurance and so avoided the loss of privacy and dignity to which most NHS patients are subjected. I could have a room to myself, with a carpet and with my own loo – details that are very important to patients but not to NHS administrators and architects. Nor, I am afraid to say, do many doctors care about these things, until they become patients and come to understand that patients in NHS hospitals rarely get peace, rest or quiet and never a good night's sleep.

I was anaesthetized and woke up a few hours later back in my room with a bandage over my eye, completely pain-free. I spent the evening drifting in and out of sleep, watching a fascinating light show in my blinded left eye, enhanced by morphine. It was as though I was flying over a pitch-black desert at night with brilliant fires burning in the distance. It reminded me of watching bush-fires at night when I had worked as a teacher in West Africa many years earlier – long walls of flame driven across the savannah grasslands by the *harmattan* wind off the Sahara, burning on the horizon beneath the stars.

The surgeon came to see me very early next morning, on his way to his NHS hospital. He took me round to the treatment room and removed the bandage from my left eye. All I could see with it was a vague dark blur – a little like being underwater.

'Bend forward and hold your watch close up to your left eye,' he said. 'Can you see anything?'

The face of my watch, hugely magnified, like the moon rising over the sea at night, swam into view.

'Yes,' I told him.

'Good,' he said cheerfully. 'You can still see.'

I was effectively blind in my left eye for the next few weeks. The gas bubble in my eye was at first like the horizon of a great planet over which I could only see a thin glimpse of the outside world. It gradually shrank and vision slowly returned – the inside of my eye was like one of those gaudy lava lamps, the bubble slowly rolling and bouncing whenever I moved my head. I was unable to operate for a month but, rather reluctantly, started doing outpatient clinics a week after my own operation. I found it quite tiring. I wore a black patch over my eye which gave me a nicely piratical look although I felt a little embarrassed that my patients could see that I was not in perfect health. When I went to see the eye surgeon a few days after the operation, sporting my eye patch, he looked dubiously at me.

'Drama queen,' he said, but otherwise reported himself

to be happy with the state of my eyeball.

I had fully recovered within a matter of weeks but one of the consequences of a vitrectomy is that the lens in the eye becomes progressively damaged and needs to be replaced. This is a simple, straightforward operation, more commonly carried out for cataracts, which I underwent three months later. I was on call for emergencies for the weekend after that second, minor operation.

If it had not been raining on the Sunday afternoon perhaps I would not have fallen down the staircase and broken my leg. Perhaps my eyesight was still out of true. After a busy Saturday night, Sunday morning was quiet. I had had to go in to operate at midnight since the on-call registrar was new and had needed help with a relatively simple operation on a middle-aged man with a cerebellar stroke. The operation had gone easily enough and I spent Sunday morning, feeling rather tired, working in my small, overgrown and ramshackle back garden.

I had driven to the Wandsworth recycling dump with plastic bags full of garden refuse, and joined the queue of polished estate cars and SUVs waiting to take part in this Sunday morning ritual. Into the huge containers of the dump people were busy throwing rubbish – the future archaeology of our civilization – broken armchairs, sofas, washing machines, hi-fi equipment, cardboard boxes, beds and mattresses, last year's lawnmowers, pushchairs, computers, televisions, bedside lamps, magazines, plasterboard fragments and rubble. There is a furtive, guilty air to these places – people avoid each other's eyes, like men in a public toilet, and hurry to get back into the privacy and luxury of their shiny cars and drive away. Whenever I go to the dump I always leave with a great sense of relief, and on this occasion decided to reward myself with a visit to the local garden centre on the way home. It was while I was walking happily between the rows of plants and shrubs, looking for something to buy, that it started to rain. A low ragged rain cloud, looking like ink spreading in clear water, raced overhead and the rain poured down, driving the shoppers indoors and leaving the garden centre suddenly deserted. I found myself standing alone among the green plants and shrubs. My mobile phone rang. It was Rob, the on-call registrar from the hospital.

'I'm very sorry to disturb you,' he said, going through the usual polite litany my juniors always recite whenever they call me, 'but could I discuss a case with you, please?'

'Yes, yes, of course,' I said, hurrying off to find shelter in a warehouse full of terracotta pots.

'This is a thirty-four-year-old man who fell from a bridge ...'

'A jumper?'

'Yes. Apparently he'd been depressed for some time.'

I asked if he had landed on his head or on his feet. If they hit the ground feet first they fracture their feet and spines and end up paralysed and if they hit their head first they usually die.

'He landed on his feet but he hit his head as well,' came the reply. 'He's a polytrauma case – he's got a fractured pelvis, bilateral tib and fibs and a severe head injury.'

'What does the scan show?'

'A large haemorrhagic contusion in the left temporal lobe and the basal cisterns are gone. He's had a big, fixed pupil on the left for five hours now.'

'And his motor response?'

'None, according to the ambulance men.'

'Well, what do you want to do?'

Rob hesitated, reluctant to commit himself.

'Well, I suppose we could pressure monitor him.'

'What do you think is his prognosis?'

'Not very good.'

I told Rob that it would be better to let him die. He would probably die whatever we did, and even if he did survive he'd be left terribly disabled. I asked him if he had seen the family.

'No, but they're coming in,' he replied.

'Well, spell it out to them,' I said.

While we were speaking, the rain had stopped and the sun had come out from behind the broken clouds. The plants around me glittered with reflected light. The shoppers emerged from the shelter of the shop and the pastoral scene of the garden centre resumed – happy gardeners walked between the rows of plants and trees, stopping to examine them, and wondering which to buy. I bought myself a *Viburnum paniculata* with little starbursts of white flowers and drove home with it perched in a friendly sort of way on the passenger seat beside me.

I could have operated on this poor suicidal man and possibly saved his life, but at what cost? Or so I told myself as I started to dig a hole in the back garden for the viburnum. Eventually I felt forced to go in to the hospital to look at the scan myself and to see the patient – despite my best efforts I found it difficult to deliver a death sentence, even on a jumper, on the basis of hearsay evidence alone.

My shoes had become soaking wet in the downpour and I changed them for a new pair, recently re-soled, before driving in to the hospital.

I met Rob in the dark X-ray viewing room. He summoned up a brain scan on the computer screens.

'Well,' I said as I looked at the CT scans, 'he's wrecked.' It was a relief that the scan looked even worse than Rob's description of it over the phone. The left side of the man's brain was smashed beyond repair, his brain darkened on the scan by oedema and flecked with white, the colour of blood on CT brain scans. His brain was so swollen that there was no hope of survival, even in a disabled state, even if we operated.

'There are two great benefits to medicine as a career,' I said to Rob. 'One is that one acquires an endless fund of anecdotes, some funny, many terrible.'

I told him about a jumper I had treated years ago, a pretty young woman in her twenties who jumped under a tube train. 'She had to have a hindquarter amputation of one leg – the leg removed completely at the level of the pelvis – I suppose the train had run lengthwise over her hip and leg. She'd also suffered a compound depressed skull fracture which was why she was sent over to us after the local hospital had done the amputation. We sorted her head out and she slowly woke up over the next few days. I remember telling her she'd lost her leg and she said "Oh dear. It doesn't sound very nice, does it?" But she was quite happy at first, obviously couldn't remember all the unhappiness that had made her throw herself under the train. But as she recovered from her head injury, as she got better, so to speak, she got worse since her memory started to return and every day you could see her become more and more depressed and desperate. When her parents eventually turned up you could see why she had tried to kill herself. It was very sad to watch.'

'What happened to her?' Rob asked.

'I haven't a clue. We sent her back to the local hospital and I heard no more.'

'What is the second benefit of a career in medicine?' Rob asked politely.

'Oh, just that if one falls ill oneself one knows how to get the best care.' I waved at the brain scan on the monitor in front of us. 'I'll go and talk to his parents.'

I left the X-ray viewing area and walked along the dull and overlit hospital corridor to the ITU. The hospital was still very new. It felt like a high-security prison – doors could only be opened with a swipe card and if the doors were left open for more than a minute an ear-splitting alarm would sound. Fortunately since then most of the alarms have broken or been sabotaged, but our first few months in the new building were spent with the almost constant sound of alarms going off – an odd phenomenon for a hospital full of sick people, one might think. I walked into the ITU. Lined around its walls were the forms of the unconscious patients on ventilators surrounded by machinery, with a nurse at each bed.

The nurses at the central desk pointed to one of the beds when I asked about the new admission and I walked over to it. I was taken aback by the fact that the poor jumper was immensely fat. For some reason I didn't expect a suicide to be fat, so fat that from the end of the bed I could not see his head at all – only the great pale mound of his naked belly, partly covered by a clean sheet, and beyond it the monitors and machinery and syringe drivers at the head of the bed, with their flashing red LEDs and digital read-outs. An elderly man was sitting on a chair at the bedside and got up when he saw me. I introduced myself and we shook hands.

'Are you his father?' I asked.

'Yes,' he said quietly.

'I'm very sorry,' I said, 'but there's nothing we can do to help.' I explained that his son would die within the next twenty-four hours. The old man said nothing other than to nod his head. There was little expression on his face – whether he was too stunned, or too estranged, I do not know. I never got to see his son's face, and I do not know what human tragedy lay behind the pathetic, dying bulk that lay on the hospital bed beside us.

I went home and climbed up the stairs to the attic room I had built the previous year to where Kate was lying on a sofa, recovering from a particularly bad relapse of her Crohn's disease. I had made the oak staircase myself and had sanded and polished the steps to a high finish. We discussed the need for an extra handrail on the stairs since Kate had slipped on them and bruised herself quite badly two nights earlier. We have both always been a little dismissive of the Health and Safety culture that increasingly dominates our risk-averse society but decided that a handrail was probably a good idea. I set off downstairs, down the handmade oak stairs, each tread and riser carefully made by me, to finish planting the viburnum in the back garden. My newly soled shoes slipped on the over-polished oak, I lost my balance, heard the horrible, explosive crack of my leg breaking and my foot dislocating and fell down the stairs.

Although breaking one's leg is indeed very painful it is surprisingly easy to tolerate – it is well known, after all, that soldiers in battle rarely feel great pain if they are seriously wounded – the pain comes later. You're too busy working out how to save yourself to think much about the pain.

'Bloody hell! I've broken my leg,' I shouted. Kate at first thought I was joking until she found me at the bottom of the stairs, with my left foot twisted round at an improbable angle. I tried to pull my foot out straight with my hands but started to pass out with the pain so Kate called our neighbours who put me on the back seat of their car and took me to A&E at my own hospital. A wheelchair was found and soon I was in a short queue at the reception desk, manned by two fierce-looking women behind what looked like bullet-proof glass. I sat there patiently, gritting my teeth, my broken leg sticking out in front of me. After a short delay I was facing one of the receptionists.

'Name?' she asked.

'Henry Marsh'.

'Date of birth?'

'Five three fifty. Actually I'm the senior consultant neuro-surgeon at this hospital.'

'Religion?' she asked in reply, without batting an eyelid.

'None,' I replied, crestfallen but thinking that at least my hospital was truly egalitarian.

The interrogation went on for a short time and I was then rescued by one of the Casualty sisters who promptly established that my foot was dislocated and that it needed reducing. I was very gratified at how quickly this was done, and painlessly at that, thanks to IV morphine and midazolam and Entonox. My last memory, before the drugs rendered me unaware of everything around me is of my trying to persuade the enthusiastic sister not to take an enormous pair of scissors to my brand new, green corduroy trousers.

When I started to come round in a happy haze from the drugs, and reflected on what it would have been like to have had a fracture like mine reduced in the past, without any anaesthetic at all, I found my orthopaedic colleague standing at the end of my trolley. I had called him on my mobile from the backseat of my neighbours' car on the way to A&E.

'It's a fracture dislocation,' he said 'They've reduced it nicely but it will need an operation – internal fixation. I could do that tomorrow at the private hospital.'

'I've got insurance,' I said. 'Yes, let's do that.'

'We'll have to get a private ambulance,' said the sister.

'Don't worry,' said my colleague, 'I can take him myself.'

So I was wheeled out of A&E, my left leg in a long plaster back-slab and helped into my colleague's red Mercedes sports car. Thus I was taken in some style to the private hospital three miles away, where the fracture was duly fixed the next day. My colleague insisted on keeping me in hospital for five days on the grounds that I was a doctor and would not listen to his medical advice that I should rest my leg for the first few days after surgery. So I spent much of the following week in bed, with my leg propped up in the air, looking at a rather fine oak tree outside the window of my room, reading P.G. Wodehouse and reflecting on the way in which many of the government's so-called 'market-driven reforms' of the NHS seemed to be driving the NHS even further away from what went on in the real market of the private sector, in which I was once again a patient. I could occasionally hear my colleagues going to visit their patients in the rooms next to mine, their voices full of charm and polite encouragement.

On the morning of my discharge I went down to the out-patient area to wait to have the plaster changed. I watched the many outpatients coming and going.

My colleagues, in smart dark suits, would emerge from time to time from their consulting rooms to bring in the next patient waiting to see them. Some of them knew me and looked somewhat startled to see me disguised as a patient in a dressing gown with a leg in plaster. Most of them stopped and commiserated and laughed with me at my bad luck. One of them, a particularly pompous physician, stopped for a moment and looked surprised.

'Fracture dislocation of the left ankle,' I said.

'Oh dear,' he said in a very prim voice, as though he disapproved of the vulgar way in which, by allowing my leg to be broken, I had become a mere patient, and he quickly returned to his room. I was summoned to the plaster room where my orthopaedic colleague removed the old dressing and carefully studied the two incisions, one on either side of my ankle. He declared himself happy and then, taking my leg in his hands, re-dressed the wounds and placed a new plaster back-slab against my foot and leg, holding it in place with a crepe bandage. I thought rather wistfully of the huge gulf that separates this sort of medicine from what I practice as a neurosurgeon.

'I rarely touch my patients, you know,' I said to him.

'Other than when operating on them, of course. It's all just the history and the brain scan and long, depressing conversations. Not at all like this. This is rather nice.'

'Yes, neurosurgery is all doom and gloom.'

'But our occasional triumphs are all the greater as a result ...' I started to say before he interrupted my philosophizing.

'You have got to keep that foot up for the next few weeks ninety-five per cent of the time because it's going to become very swollen.'

I bade him goodbye and, picking up my crutches, hopped out of the room.

A few weeks later I suffered a vitreous haemorrhage and retinal tear in my other eye but it was easier to fix than the left eye had been. I was back at work within a matter of days. I had been lucky compared to my patients, and I was full of the profound and slightly irrational gratitude for my colleagues that all patients have for their doctors when things go well.

22

ASTROCYTOMA

n. a brain tumour derived from non-nervous cells. All grades of malignancy occur.

After the success of the trigeminal neuralgia operation Igor was keen that when I next went out to Ukraine I should operate on a number of patients with especially challenging brain tumours, which he assured me could not be treated safely in Ukraine by his senior colleagues. I did not share his enthusiasm and told him so, but when I arrived on my next visit there was a long queue of patients with quite dreadful brain tumours waiting to see me in the dingy corridor outside his office.

The outpatient clinics I have conducted over the years in Igor's office have always been bizarre events, and quite unlike anything else I have ever done. As Igor's fame grew, patients would come from all over Ukraine to consult him. There was no appointment system – patients would turn up at any time, and seemed to accept that this might involve waiting all day to be seen. On the occasion of my visits the queue of patients would stretch all the way down the long hospital corridor outside his office and disappear from view around a distant corner.

We would start at eight in the morning and continue until late in the evening without a break. There would often be several patients and their families in the small office all at once, some of them dressed, some of them undressed. There might also be journalists and TV crews conducting interviews, especially when Igor's political situation was proving problematical. There were three telephones in the room and most of them were in constant use. There were rarely fewer than seven or eight people in the room at the same time. I found all this chaos exhausting and irritating and at first I blamed Igor for it, telling him that he should institute an appointments system, but he said that in Ukraine nobody would adhere to it and that it was best to let people turn up whenever they wished.

Igor's manner with patients was somewhat brusque, although at times he seemed capable of sympathy. Since I do not speak Russian or Ukrainian I could only guess at what was being said, before it was translated by Igor, and I discovered that often I was entirely wrong. The patients brought their own brain scans, arranged previously, and without further ado I would be asked whether surgery was possible or not. In English medicine it is drilled into one at a very early stage that one should base one's decisions on taking a history and examining the patient and that only at the end should one look at the 'special investigations' such as X-rays and brain scans. Here the whole process was reversed and compressed into a few minutes or even seconds. I felt like the emperor Nero at the Roman games. It was made all the more difficult by the fact that the brain scans were usually of poor quality. It was difficult to see clearly what was going on and this made me even more uneasy about having to make so many rapid life-or-death decisions.

On this particular visit, in the summer of 1998, it became clear that Igor's many enemies in the medical establishment had brought pressure to bear on the hospital director who had welcomed the British ambassador the previous year. On the morning of the first outpatient clinic I learnt that I had been 'banned' from the operating theatres by the director, and also that he would not meet me. I was, in fact, quite relieved – the cases I had seen were daunting and I was frightened by the thought of operating on them in the primitive operating theatres.

The fact that I had been banned from the operating theatres was headline news and the outpatient clinic next day had more than its usual share of journalists and TV crews in attendance. Halfway through the morning, as I was being interviewed by a Ukrainian TV journalist, while simultaneously trying to decide whether somebody's brain tumour was operable or not, the head of the hospital's surgical department arrived and ordered the journalists and film crews out of the hospital. He wore a particularly tall chef's hat, and an outsize pair of spectacles to go with it, and looked reassuringly absurd. It was difficult to take him seriously. We left the hospital and continued the interview outside, with the hospital in the background.

One of the patients I had just seen, and agreed – with considerable misgivings – to operate on, was also interviewed and asked what she felt about the fact that I was not going to be allowed to treat her. Ludmilla had come up from the south of the country to see a famous professor of neurosurgery in Kiev. She had become increasingly unsteady on her feet in recent months and a brain scan had shown a large and very difficult tumour at the base of her brain – an ependymoma of the fourth ventricle, a benign but often fatal growth. There was no question of her undergoing surgery in her home town. She arrived on time for her appointment but the professor was late. His junior doctors looked at her brain scan.

'If you want to live, leave before the professor returns,' one of them had said. 'Go and see Kurilets. He has contacts with the West and may be able to help. If you let the professor operate, you will die.' She quickly left and a few days later I saw her in Igor's office.

We both appeared on the national nine o'clock television news that evening.

'What do you want?' the journalist was seen to ask Ludmilla.

'I want to live,' she replied quietly.

The urge to help, the planning of difficult and dangerous operations, of taking carefully calculated risks, of saving lives, is irresistible and even more so if you are doing it in the face of opposition from a self-important professor. When I met Ludmilla the next day I felt that I had no choice other than to tell her that, if she wished, I would arrange for her to come to London and I would operate on her there. Not surprisingly, she agreed.

It was the next day that I first saw Tanya. Igor wanted us to leave for the hospital by 6.30 in the morning but I overslept and once we had set off I quickly realized why Igor had been keen to leave early – the morning Kiev rush hour meant that a journey of thirty minutes took, instead, one and a half hours. We joined an endless queue of grubby cars and trucks, dull grey shapes in the fog, with red tail-lights turning their exhaust into small pink clouds, inching along the enormous wide roads towards the centre of Kiev. The roads are lined by huge advertisements for cigarettes and mobile phones, scarcely visible in the fog. Many cars queue-jump by mounting the pavements and weaving between the lamp posts. Heavy 4x4s leave the road altogether and career across the muddy patches of grass beside the road if it will get them ahead.

Tanya was near the end of the queue of patients with inoperable brain tumours. She was eleven years old at the time. She walked into Igor's office, unsteadily, supported by her mother, with a scratched piece of X-ray film that showed an enormous tumour at the base of her brain that must have been growing for years. It was the largest tumour of its kind that I had ever seen. Her mother, Katya, had brought her all the way from Horodok, a remote town near the border with Romania. She was a sweet child, with the awkward long-legged grace of a foal, a page-boy haircut and a shy, lopsided smile – lopsided because of the partial paralysis of her face caused by the tumour. The tumour had been effectively deemed inoperable both in Moscow and Kiev and it was obvious that it was going to kill her sooner or later.

Just as it is irresistible to save a life, it is also very difficult to tell somebody that I cannot save them, especially if the patient is a sick child with desperate parents. The problem is made all the greater if I am not entirely certain. Few people outside medicine realize that what tortures doctors most is uncertainty, rather than the fact they often deal with people who are suffering or who are about to die. It is easy enough to let somebody die if one knows beyond doubt that they cannot be saved – if one is a decent doctor one will be sympathetic, but the situation is clear. This is life, and we all have to die sooner or later. It is when I do not know for certain whether I can help or not, or should help or not, that things become so difficult. Tanya's tumour was indeed the largest I had ever seen. It was almost certainly benign and, at least in theory, could be removed, but I had not tried to operate on such a large tumour in a child her age before, nor did I know anybody else who had tried. Doctors often console each other, when things have gone badly, that it is easy to be wise in retrospect. I should have left Tanya in Ukraine. I should have told her mother to take her back to Horodok, but instead I brought her to London.

Later that year I arranged for Tanya and Ludmilla to come to London and I organized a minivan to meet them at

Heathrow and deliver them with their accompanying relatives at the entrance to my hospital. How proud and important I felt when I met them there! I carried out both the operations with Richard Hatfield, a colleague and close friend who had often come out to Ukraine with me.

The operation on Ludmilla took eight hours and was a great success. The first operation on Tanya took ten hours, and then there was a second operation that took twelve hours. Both operations were complicated by terrible blood loss. With the first operation she lost four times her entire circulating blood volume but she emerged unscathed, although with half of the tumour still in place. The second operation – to remove the rest of the tumour – was not a success. She suffered a severe stroke. She had to stay in the hospital for six months before she was, more or less, well enough to return home to Ukraine. I drove her and her mother to Gatwick, with the help of Gail and her husband. We stood by the departures gateway. Tanya's mother, Katya, and I kept on looking each other in the eyes – she with desperation, I with sadness. We embraced, both of us crying. As she started to wheel Tanya in her wheelchair through the gate she ran back to me and hugged me again. And so they left – Katya pushing her mute and disfigured daughter in her wheelchair, and the Ukrainian doctor Dmitri beside them. Katya probably understood better than I did what the future would bring.

Tanya died eighteen months after her return home. She would have been just twelve years old. Instead of a single, brilliant operation she ended up having to undergo many operations and there were serious complications – 'complications' being the all-encompassing medical euphemism for things going wrong. Instead of a few weeks she had ended up spending six months in my hospital, six horrible months. Although she did eventually get home she returned more disabled than when she had left it. I don't know exactly when she died and I only got to hear of it by chance from Igor. I had telephoned him from London to discuss another brain tumour case. I had asked in passing, a little anxiously, after Tanya.

'Oh. She died,' he said. He didn't sound very interested. I thought of all that Tanya and Katya had been through, of what we had *all* been through in our disastrous efforts to save Tanya's life. I was upset, but his spoken English is limited and broken and perhaps something was being lost in translation.

I had last seen her shortly before her death, after her return from England, on one of my regular trips to Kiev. Katya, her mother, had brought her to see me all the way from her home in Horodok. She could just walk if somebody held her, and her faint, lopsided smile had returned. For the first few months after the operations her face had been completely paralysed. This had left her at first not just unable to talk but also with a face like a mask, so that it seemed she had no feelings at all – even the most intense emotions were hidden, unless sometimes a tear rolled down her expressionless cheek. It is sad how easy it is to dismiss people with damaged or disfigured faces, to forget that the feelings behind their mask-like faces are no less intense than our own. Even then, a year after surgery, she was still unable to talk or to swallow although she could now breathe without a tracheostomy tube in her throat. Katya had been with her in London throughout those endless six months, and when I had seen them off at Gatwick Airport, Katya had sworn to give me a present whenever we next met. She now came not just with Tanya but with a large suitcase. This contained the family pig, which had been slaughtered in my honour and turned into dozens of long sausages.

A few months later Tanya was dead. She probably had died from a blocked shunt. After the catastrophic second

operation on her brain tumour I had had to insert an artificial drainage tube into her brain and this might well have
blocked, causing a fatal increase in the pressure inside her
head. Living, as she did, far away from modern medical facilities, it would have been impossible to deal with this. I will
never know for certain what actually happened. Nor will I
ever know whether I had been right to uproot her from her
home in impoverished, rural Ukraine for so many months,
and to operate on her in the way that I had done. For the
first few years after Tanya's death Katya would send me a
Christmas card – coming all the way from Horodok it did
not usually reach me until the end of January. I would put
it up on my desk in my windowless office in the huge, factory-like hospital where I work. I would leave it there for a
few weeks as a sad reminder of Tanya, of surgical ambition
and of my failure.

Several years after Tanya's death a documentary film about
my work in Ukraine was being made and I suggested that it
should include a visit to see Katya. The film crew and I were
driven in a minibus the four hundred kilometres from Kiev
to Horodok. It was in late winter and much of the filming
had gone on in deep snow and in temperatures of minus 17
degrees Celsius but as we drove west the snow faded away
and, although all the rivers and lakes we passed were frozen
solid, often with men fishing through holes sawn in the ice,
there was a distinct feeling of spring in the air. I wanted very
much to see Katya again – during the six months she and
Tanya had been in London I had come to feel very close to
them, even though we had no language in common. I also
felt very anxious as I could not help but blame myself for
Tanya's death.

   Horodok, as with so much of rural western Ukraine, was
impoverished and depopulated. Since the fall of the Soviet

Union the economy had collapsed and most of the young people had left. There were the rust-coloured, derelict factories that are to be found all over Ukraine and rubbish and broken machinery were scattered everywhere. Katya lived in a small brick-built house beside a muddy yard – when we arrived she seemed as nervous as I was, although she was clearly very happy to see me. We waded across the mud and puddles to reach the little house where there was an enormous meal laid out for us. We sat down with her family round the table as the film crew filmed us. I was so intensely moved to see Katya again that I could scarcely talk, and was quite unable to eat, much to Katya's distress. I managed to stumble through giving a toast as we followed the Ukrainian tradition of proposing toasts with vodka, accompanied by short speeches.

The following day we went to visit Tanya's grave in the local cemetery. It was some miles away from Katya's house, standing on its own beside a wood. The road to it wound along country lanes, lined with bare winter trees, passing through battered and dishevelled villages, each with a pond, frozen over with blue-grey ice, with geese and ducks standing at the edges. Orthodox cemeteries are wonderful places. The graves are decorated with dozens of brilliantly coloured artificial flowers and the gravestones all have photographs behind glass or portraits etched in stone of the deceased. Everything was in perfect order and in marked contrast to the dilapidated houses in the villages of the living which we had passed through on our way to the cemetery.

Tanya's grave had a six-foot-high headstone from which her carved face appeared – odd, perhaps to western eyes, but beautiful. The sun was shining, the artificial flowers glittered and shook in a light wind and in the distance I could hear the chickens in the local village. The snow had melted and only a little was left, showing up as white lines in the furrows of

the ploughed field which we had crossed to reach the cemetery. There was birdsong everywhere. As the film crew set up
their equipment I wandered round the cemetery, looking at
the gravestones and their portraits. Most of the people buried
here would have lived through the most terrible times – the
Civil War of the 1920s, the famine of the 1930s (though it
had been worst in central Ukraine), Stalin's despotism and
the unspeakable horrors of the Second World War. At least
a quarter of the population of Ukraine died violently in the
twentieth century. I wanted to ask these dead faces what they
had done during those years, and what compromises they
must have made to survive, but it seemed to me that they
looked back at me as though to say: 'We are dead. You are
still alive. And what are *you* doing with the time that you
have left?'

The film about Igor and me was a great success. It has
been shown all over the world and won many awards. At
the end of the film I was shown standing in front of Tanya's grave. I was looking sad not just because of Tanya's
death but because next to her grave, and unbeknown to any
viewers, was her father's grave. He had gone to Poland a
few months earlier to make some money as an agricultural
labourer since he and Katya were desperately poor. He had
managed to earn a thousand dollars and was about to set
out for home for Christmas when he was found murdered.
The money had gone. I had wanted to see Katya not just
because of Tanya but also because of her father's death. Life
in Ukraine is not easy.

23

TYROSINE KINASE

n. an enzyme that acts as an on/off switch in many cellular functions. Drugs to reduce its activity, known as tyrosine kinase inhibitors or TKIs, are used in the treatment of many cancers.

'Are we quorate?' the chairman asked. A rapid head count showed that we were, so the meeting began.

The chairman, after making a few brief jokes, got to the business of the meeting.

'We have patient representatives from the Support Group for the technology we are discussing today,' he said, looking towards three elderly grey-haired men who sat on one side of the hollow square of tables around which the Technology Appraisal Committee was seated. 'Welcome!' he said, with an encouraging smile. 'We have our clinical experts,' – he pointed to two grave-looking men next to the patient representatives – 'and we have representatives from the company whose drug for this cancer we are considering,' he continued, in a slightly more formal voice, looking towards two anonymous-looking men in dark suits with large box-files on the floor in front of them. They sat a few feet behind us, away from the tables.

'Mr Marsh is the Clinical Lead and will tell us about the evidence for the effectiveness of the drug, but I thought we might first start with statements from the patient representatives.'

The first of the three elderly men cleared his throat a little nervously and, with a sad and resigned expression, delivered his statement.

'I was diagnosed with the cancer two years ago,' he began, 'and at the moment am in remission. I have been told that it's bound to start growing sooner or later and the only possible treatment when that happens will be this new drug you are considering today ...'

As he spoke, the committee listened in complete silence. It was difficult not to admire his bravery in talking to a room full of strangers in this way. He went on to explain that he had started a support group for patients with this particular disease.

'There were thirty-six of us to begin with but now there are only nineteen of us left. I would ask you to remember when you consider this drug,' he added, with a slight note of despair as he finished, 'that life is precious, that every day counts ...'

The next elderly man spoke of how his wife had died from the cancer, and he told us of her suffering and the misery of her final months. The third elderly man opened the briefcase in front of him and pulled out a sheaf of papers. He looked very determined.

'I am only here,' he began, 'in my opinion, because of this drug. I was first diagnosed twelve years ago – and as you all know most people die from it in less than five years. The doctors here had nothing to offer me so I read up about it and went to America and was enrolled in various drug trials. The last drug was the one you are looking at today – I started it eight years ago. The NHS would not give it to me. It has cost me three hundred thousand pounds of my own money so far. Gentlemen ...' he looked around the room at us all, 'I hope you will not consider me to be a statistical outlier.'

After a brief pause the chairman turned to me. 'Mr Marsh will now tell us about the clinical effectiveness of the drug in question.' He pushed the laptop in front of him towards me.

I had volunteered my services to NICE, the National Institute of Clinical Excellence, two years earlier. I had seen an advertisement in one of the medical journals for a consultant surgeon to join one of the NICE Technology Appraisal Committees. I thought the word 'technology' would mean interesting things like microscopes and operating instruments but it turned out, to my dismay, to mean drugs. The only exam I ever failed in my prolonged academic career was in pharmacology. The popular press often accuses NICE of being an organization of callous bureaucrats – in America right-wing politicians refer to it as a 'Death Panel'. These are wholly unfair accusations and as I have become familiar with the process by which new drugs are appraised by the committee, and a decision made as to whether the NHS should use them or not, I have become increasingly fascinated. Once a month I will take the train to Manchester, where the all-day meeting is held in the NICE head office. The members take it in turns to present the evidence about the drugs being considered. On this occasion it was my turn.

The slides for my presentation were projected, one by one, on three of the room's four walls as I spoke. They were rather dull slides, plain blue letters on a white background, with facts and figures and the long unpronounceable names of chemotherapy drugs, over which I stumbled as I read. I had prepared the slides in a frantic rush with the help of the NICE staff over the preceding few days. The meetings are open to the public and there can be none of the jokes and pictures trawled off Google Images with which I usually decorate my medical lectures. My presentation took about ten minutes.

'The conclusion,' I said as I finished, 'is that this TKI works for this particular cancer in the sense that it significantly reduces the size of the patient's spleens but this is only a surrogate outcome. It is not clear from the trials whether patients lived longer or had a better quality of life. Many of the patients were lost to follow-up and the quality of life data is largely missing.'

There was then a ten-minute break for coffee. I found myself standing next to the chairman. I told him that I had been in Ukraine two weeks earlier and had been told that drug trials there are a good little money-spinner. Many of the hospitals are involved in trials for the big drug companies and I was told that the same patient might be put into several different trials since the doctors get paid for every patient they enter. If that is true, I said, the results are therefore meaningless. The chairman chose not to comment.

The next presentation was by a health statistician and dealt with the cost-effectiveness of the drug – in other words the question of whether the benefits for patients dying from the cancer are worth the drug's cost. He had the hesitant delivery of an academic and he stumbled and hesitated as he went through his complex slides. His presentation was a series of graphs and tables and acronyms, using the various models that health economists have developed in recent years to analyse this question. I quickly became lost and furtively looked around me, trying to guess if the other committee members understood his presentation any better than I did. They were not giving anything away and were all watching the projection screens with expressionless faces.

In this kind of economic evaluation the extra life that patients may, or may not, get from a drug is adjusted to make allowances for the fact that the extra time might, or might not, be of only poor quality. Most patients dying from lung cancer, for instance, will be in poor health – short of breath, coughing blood, in pain, in fear of death. If they were to live an extra year (which is unlikely with that particular cancer once it has spread) and were in good health, that extra year would be given the value of one year. If they were in poor health, that value would be correspondingly reduced. This value is called a Quality Adjusted Life Year and it is calculated using 'utilities'. In theory this involves asking dying patients how they feel about the quality of their life, but it has proved very difficult to do this in practice since it often involves openly confronting dying patients with their imminent death. Not surprisingly this is something from which both doctors and patients shy away. Instead, healthy people are asked to imagine that they are dying, that they are coughing blood or in pain, and then asked how much they feel this would reduce the quality of their life. Their replies are used to calculate the quality of the extra life gained by using the new cancer drug. There are various ways of doing this – one is based on a technique from game theory called the 'standard gamble'. It was invented by the great mathematician von Neumann who – it is perhaps worth pointing out – also recommended on the basis of game theory a pre-emptive nuclear strike against the Soviet Union in the days of the Cold War. Some might conclude that the standard gamble is not necessarily the best basis for human decision-making.

The degree of uncertainty that surrounds all these calculations must also be measured, which makes matters even more complicated. At the end of all this, a final figure is generated – the Incremental Cost Effectiveness Ratio, which is the cost of one extra quality-adjusted life year which the new treatment achieves when compared to the best current alternative. If this is more than £30,000 NICE will not approve the use of the drug by the NHS, although exceptions will sometimes be made for patients dying from rare cancers. Whenever NICE refuses to approve a drug there is an inevitable outcry from patient groups and the drug companies. Patients dying from various distressing diseases will appear on the television news accusing the NHS and NICE of abandoning them. NICE will be accused of being a Death Panel.

The health economist, who looked more like a harmless drudge than a sinister death panellist, trudged through his complex slides. The talk seemed to consist of nothing but acronyms and I constantly had to ask the friendly analyst next to me what they meant. Once he had finished, the chairman of the committee asked the visiting experts for their opinion and the committee members then questioned them.

'How can we judge the value of the drug if the trials don't really tell us how the patients were doing and only how long they lived?' I asked.

There was a grave and bearded professor of oncology attending the meeting as an expert witness.

'If you look at the Manufacturer's Submission,' he said in a very soft voice, which I could scarcely hear, 'you will see that the quality of life data wasn't collected because the clinicians running the trial felt it would be bad for the patient's welfare. It's a standard problem with cancer chemotherapy trials – it's difficult to get dying patients to complete questionnaires. One has to use standard utilities instead. But it's one of the few chemotherapy agents we have for this cancer that has very few side effects,' he added.

He spoke movingly of the difficulties of treating dying patients, and the fact that he had so few effective treatments.

'We would very much like to have the choice of using this drug,' he concluded.

'At any cost?' the chairman asked, delivering the *coup de grâce*. The expert had no answer to this terrible question. Once the discussion had ended the patient representatives and experts and outside observers were then ushered out of the room and the second part of the meeting – where a decision is made whether to allow the NHS to use the drug or not, but in camera – started.

'Surely,' I wanted to say to the hard-nosed health economists and public health doctors around me, but did not dare, 'the real utility of the drug is to give dying patients hope? The hope that they might be statistical outliers and live longer than average? How do you measure the utility of hope?'

I could have delivered an impassioned lecture on the subject. I have spent much time talking to people whose life was coming to an end. Healthy people, I have concluded, including myself, do not understand how everything changes once you have been diagnosed with a fatal illness. How you cling to hope, however false, however slight, and how reluctant most doctors are to deprive patients of that fragile beam of light in so much darkness. Indeed, many people develop what psychiatrists call 'dissociation' and a doctor can find himself talking to two people – they know that they are dying and yet still hope that they will live. I had noticed the same phenomenon with my mother during the last few days of her life. When faced by people who are dying you are no longer dealing with the rational consumers assumed by economic model-builders, if they ever existed in the first place.

Hope is beyond price and the pharmaceutical companies, which are run by businessmen not altruists, price their products accordingly.

The admirable purpose of NICE's technology appraisal (which is only one part of NICE's work) is to try to provide a countervailing force to the pharmaceutical companies' pricing policies. The methodology used for the drug in question was unrealistic, verging on the absurd, and I wondered how many of the people sitting round the hollow square

understood the difficulties and deceptions involved in treating patients who are dying, where the real value of a drug
such as this one is hope, and not the statistical probability
of living, possibly in great pain, for an average of an extra
five months.

I kept my doubts to myself since I firmly believe that the
pricing policies of the great drug companies must be resisted
and that health-care costs, like greenhouse gases, must be
curbed. The abstract discussion continued.

'But the MS doesn't even involve a PSA!' a young health
economist was saying with deep indignation. 'And if you
want my opinion we should toss this application out ...'

'Surely not Prostate Specific Antigen?' I asked my neigh-
bour, unable to resist a silly joke.

'No,' he said. 'Probabilistic Sensitivity Analysis.'

'Well, I have some problems with PSAs,' the chairman
said, 'but the assumptions about the *haq* slope are important and the lowest possible *icer* is £150,000 so even though
EOL applies there is no way this drug will pass. At a cost of
£40,000 per year for treatment per patient there was never
any chance it could be cost-effective.'

This last one at least was an acronym I knew – End of
Life was a compromise NICE had recently been forced to
make to allow the use of expensive drugs in small groups of
patients dying from rare cancers.

The discussion went on interminably. Half of my fellow
committee members spoke and argued in the arcane language
of cost-effectiveness analysis with passion and assurance,
while the other half nodded wisely.

Did they really understand all this? I felt ashamed by my
ignorance.

Eventually the chairman looked round the committee.

'I think we are looking at a minded "No" here, are we
not?' he said.

What this means is that the committee's recommendation will go out to consultation and all the interested parties – patient groups, manufacturers, clinicians – can criticize and comment before a final conclusion is reached. NICE bends over backwards to be transparent and to involve all the 'stakeholders' involved in its deliberations and its portrayal in the media. Besides, it is just possible that the pharmaceutical company who make the drug might reduce the price.

I took the train back to London that afternoon and arrived back at Euston at seven in the evening. I walked the two miles to Waterloo in the dark January evening and along with hundreds of commuters crossed the bridge over the oil-black river, the city to either side of us made wonderful by millions of electric lights glittering in the night over snow-covered roofs. It was good to escape, if only for a few hours, the world of disease and death in which I spend so much of my life.

24

OLIGODENDROGLIOMA

n. a tumour of the central nervous system.

It was Sunday evening, and there were three patients with brain tumours on the operating list for the next day: a woman my own age with a slowly growing meningioma, a young doctor with an oligodendroglioma that I had already operated on some years earlier which was now growing back again and which we both knew would ultimately prove fatal, and an emergency admission whom I had not yet seen. I took my bicycle to the hospital's basement entrance, next to the wheelie bins where the nurses sometimes come to smoke cigarettes. The door's lock seems to be permanently broken so I can enter the building and take my bike up in the service lift to the back of the operating theatres where I then leave it and go to see my patients. I went to the women's ward first to find the woman with a meningioma. I met one of the senior sisters – a friend for many years – coming down the corridor. She had a coat on and must have finished her shift. She was almost in tears.

I put my hand out to her.

'Oh it's hopeless,' she said. 'We've got so many staff short-ages this week and all we can get at night are agency nurses who are worse than useless. And the news has all these stories about bad nursing care ... but what can we do about it?'

I looked at the white board on the wall beside the nurses' station which lists all the patients on the ward. The patients are moved around so continually because of the lack of beds that the board is rarely up to date and finding patients is often very difficult. I could not see my patient's name on the list. There was a group of nurses who were laughing and shouting by the desk, and as far as I could tell what they were discussing had nothing to do with the patients.

'Where's Mrs Cowdrey, the patient for tomorrow's list?' I asked.

One of the agency nurses looked briefly at me and then pulled out a printed sheet of paper from her pocket which listed all the patients. She looked at it uncertainly, shrugged her shoulders and mumbled something.

'Who's in charge?' I asked

'Chris.'

'Where is she?'

'She's on a break.'

'Do you have any idea where Mrs Cowdrey might be?'

'No,' she said with a shrug.

So I walked down the corridor to the men's ward, which has some side rooms in which female patients are occasionally placed.

I found a nurse whom to my relief I did recognize – one of the many Filipino nurses in the department whose friendly and gentle nursing skills cannot be too highly praised.

'Ah! Gilbert,' I said, happy to find somebody I knew 'Have you got my woman with a meningioma for tomorrow?'

'Sorry, Mr Marsh, no. Only the two men. Maybe you should try Kent Ward?'

So I headed off up the stairs to Kent, the neurology ward. For reasons known only to themselves the management have recently rearranged our bed allocation, turning half the female neurosurgical ward into a ward for neurological stroke patients and relocating the displaced neurosurgical patients onto the neurological ward on the next floor up. So I trudged up the stairs to the neurological ward. The entrance was locked and I found that I had left my swipe card at home so I rang the bell beside the door. I had to wait many minutes before the lock buzzed and I could push the door open. I walked down the yellow-walled corridor beyond, with the patient bays to one side, each with six beds, packed closely together, like stalls in a cattle shed.

'Have you got any of my patients for surgery tomorrow?' I said hopefully to a tall male nurse sitting at the nursing station.

He looked dubiously at me.

'I'm Mr Marsh the consultant neurosurgeon,' I said, irritated at not being recognized in my own hospital.

'Bernadette's in charge. She's with a patient in the shower,' he replied in a bored voice.

So I waited until Bernadette, wearing a pair of large white wellington boots and a plastic apron, emerged from the shower room, leading a bent and elderly woman on a Zimmer frame.

'Oh, Mr Marsh!' she said with a smile. 'Are you on your usual hunt for your patients? We haven't got any here this evening.'

'This really is driving me nuts,' I said. 'I don't know why I bother. It's taken me twenty minutes just not to find one patient. Maybe she's not coming in this evening after all.'

Bernadette smiled sympathetically at me.

I found the second patient – the young doctor – sitting at one of the tables outside on the balcony between the male and female surgical wards, working on his laptop.

The original plans for this wing of the hospital – built ten years ago – had been for a bigger building than the one which was eventually built. It was built under the Private Finance

Initiative favoured by the government of the time and as with most PFI schemes the design of the building was dull and unoriginal. Nor was it cheap, since PFI has proved to be a very expensive way of building second-rate public buildings. Some would consider PFI to be an economic crime, although nobody is to be held responsible for it. It is clear now that PFI was part of the same debt-crazed culture that gave us Collateralized Debt Obligations and Credit Default Swaps and all those other dishonest acronyms and financial derivatives that have brought us (though not the bankers) to the edge of ruin.

Various parts of the design were lopped off, resulting in large and unusual balconies outside the wards. The hospital management did not see this as an opportunity for improving the patients' experience of being in hospital and instead saw it only as a suicide risk. Patients and staff were banned from the balconies and the glass doors leading onto them were kept locked. It took me many years of campaigning and large sums of charitable money raised (which then went to the private company who built and own the building) to have a small section of the balconies 'suicide-proofed' by raising the glass balustrades. I could then have the enclosed area turned into a roof garden. It has proved hugely popular with both staff and patients and on summer weekends, if the weather is fine, there is the happy sight of the ward beds almost all empty, with the patients and their families out on the balcony, surrounded by green plants and small trees, beneath large umbrellas.

This particular patient was an ophthalmic surgeon in his early forties. A gentle, mild-mannered man – which ophthalmic surgeons tend to be – he looked younger than his years and I knew that he had three young children. He worked in the North but had chosen to be treated away from his own hospital. Five years earlier he had suffered a single epileptic fit and a scan had shown a tumour growing in his brain, at the back, on the right. I had operated and removed most of the tumour but the analysis showed it to be of a type that would eventually grow back in a malignant form. He had made a good recovery but it was some time before he had regained sufficient confidence to get back to work. He had known that the tumour would recur but we had both hoped that it would take more than five years. He had radiotherapy treatment after the first operation and had remained perfectly well, but a routine follow-up scan had now shown the tumour was growing back again, and that it also now looked malignant. Further surgery might buy him a little more time, but it was unlikely to be more than five years.

I sat down beside him. He looked up from his laptop.

'Here we go again,' he said with a sad smile.

'Well, it's only a small recurrence,' I said.

'I know it can't be cured,' he said bitterly, 'but you'll take out as much as you can, won't you? This thing,' and he waved his arm towards his head, 'that is slowly doing me in.'

'Yes, of course,' I replied, handing him the consent form to sign – like all patients he scarcely glanced at it, and with a pen scrawled his name in the place I indicated. He had come to see me in my outpatient clinic some weeks earlier and we had talked over the details of the operation. We both knew what awaited him and there was nothing to say. Doctors treat each other with a certain grim sympathy. The usual rules of professional detachment and superiority have broken down and painful truth cannot be disguised. When doctors become patients they know the colleagues treating them are fallible and they can have no illusions – if the disease is a deadly one – about what awaits them. They know that bad things happen and that miracles never occur.

I cannot even begin to imagine what I would think or feel

if I knew that a malignant tumour was starting to destroy my brain.

'You're first on the list tomorrow,' I said, as I pushed my chair away and stood up. 'Eight-thirty sharp.'

Three days earlier the juniors had admitted an alcoholic man in his forties who had been found collapsed on the floor of his home, with the left side of his body paralysed. We had discussed his case at the morning meeting, in the slightly sardonic terms that surgeons often use when talking about alcoholics and drug addicts. This does not necessarily mean that we do not care for such patients, but because it is so easy to see them as being the agents of their own misfortune, we can escape the burden of feeling sympathy for them.

The brain scan had shown a haemorrhagic glioblastoma.

'See if he gets better on steroids and we can also wait and see if any family or friends turn up,' I had declared.

'His wife kicked him out of the family home some time ago,' the registrar presenting the case said. 'Because of the booze.'

'Wife beater?' somebody asked.

'I don't know.'

I found him lying sprawled on his bed, his paralysis a little better as a result of the steroids. He was a few years younger than me, overweight, with a florid red face and straggly long grey hair. I had to make a conscious decision to sit down on his bed next to him. I did not look forward to the conversation we were going to have. It is always easier to stay standing at the bedside, towering over the patient, and to leave as quickly as possible.

'Mr Mayhew,' I said, 'I'm Henry Marsh, the senior consultant. What have you been told so far about why you are here?'

'I've been told five different things,' he said desperately. 'I don't know ...' His voice was slurred by his paralysis and the left side of his face drooped lopsidedly.

'Well, what did you understand?'

'There's a tumour in my head.'

'Well, I am afraid that's true.'

'Is it cancer?'

This is always a critical point of such conversations. I have to decide whether to commit myself to a long and painful exchange, or talk in ambiguities, euphemisms and obscure technical language and leave quickly, untouched and uncontaminated by the patient's suffering and illness.

'I am afraid it probably is,' I replied.

'Am I going to die?' he shouted back in mounting panic. 'How long have I got?' He started to cry.

'Maybe you've got twelve months ...' I blurted out, instantly regretting what I had said, and alarmed at his lack of restraint. I found it very difficult to console this fat, alcoholic and pathetic man suddenly faced by his impending death. I knew that I was both awkward and inadequate.

'I'm going to die in twelve months!'

'Well, I said *maybe*. There is some hope ...'

'But you know what it is, don't you? You're the senior doc, aren't you? I'm going to die!'

'Well, I'm ninety per cent certain. But we ...' I said, lapsing into the plural form so loved by policemen and bureaucrats and doctors which absolves us from personal responsibility and relieves us of the awful burden of the first person singular, 'we might be able to help with an operation.'

He cried and cried.

'Have you got any family?' I asked, although I already knew the answer.

'I'm all alone,' he replied, through his tears.

'Any children?'

'Yes.'

'Don't they want to come and see you – even now when you're ill?' I asked and, once again, immediately regretted it.

'No.' He burst into floods of tears again. I waited for him to stop and we sat together in silence for a while.

'So you're all alone?' I said.

'Yes,' he said. 'I used to work in a hospital, you know. I'm going to die there aren't I? All the piss and shit ... All I want is a cigarette. You've just told me I'm going to die. I want a cigarette.' As he said this he desperately mimed smoking a cigarette with his one good hand, as though his life depended on it.

'You'll have to ask the nurses – nobody is supposed to smoke anywhere here at all,' I replied. I thought of all the No Smoking notices in the hospital and the huge posters that greet one on arrival at the hospital gates saying in brutal black and red, 'PUT IT OUT!'

'I'll go and talk to the nurses,' I said.

I went to find a sympathetic staff nurse.

'I've just told poor Mr Mayhew he's going to die,' I told her apologetically. 'He's dying for a cigarette. Can you help?'

She nodded quietly.

When I later left the ward, as I walked down the corridor, I saw two of the nurses lifting him into a wheelchair. He was shouting as they manhandled him off his bed.

'He's just told me I'm going to die! I'm going to die ... I don't want to die.'

There must be some secret place in the hospital where they can wheel the paralysed patients for a smoke. I was happy to know that common sense and kindness had not yet been completely beaten out of the nurses.

Three years ago I built a roof extension in the attic of my house. I installed sloping skylights with French windows opening out onto a small enclosed balcony, cut into the roof-space at the back of the house, and around this I built a short balustrade. There is room for a single chair and a few pot plants and I like to sit there in the summer evenings when I get home from work. So I sat there after returning from the hospital, with a gin and tonic, with a typical south London vista of chimney pots and slate roofs and a few tree tops, stretching away from me into the distance. I could see the garden birds flitting in the fading light between the trees of the back gardens below me, and my three bee-hives in front of my workshop. I thought of my patients. I thought of my colleague, and of the man to whom I had just read out a death sentence. I thought of how he had immediately understood that he would never get home, that his estranged family would never visit him, that he would die in the care of strangers in some impersonal place. I thought of how I had walked away – but what else could I have done? As the sun set I could hear a blackbird on the neighbouring roof singing its heart out.

The three operations, which I carried out next day, were simple and straightforward. It turned out that the woman with a meningioma had been on one of the other wards on the Sunday evening after all.

A few days later, after I had operated on the alcoholic Mr Mayhew, and after he had been discharged from my ward, I saw him in the distance as I was coming into the hospital. A nurse was pushing him in a wheelchair towards the hospital coffee shop. He waved his good arm to me and it was difficult to know if it was in greeting or in farewell. I did not see him again.

25

ANAESTHESIA DOLOROSA

n. severe spontaneous pain occurring in an anesthetic area.

It was the summer when I had broken my leg falling down the stairs. There had been a heatwave which had come to an end with a brief thunderstorm early in the morning. I lay happily in bed listening to the thunder rolling and crashing over the silent city. My plaster cast had been removed the day before and replaced by a large plastic and Velcro inflatable boot that looked as though it belonged to an Imperial Trooper from *Star Wars*. It was very clumsy, but at least I could walk again and take it off at night. It was strange to be reunited with my leg and to see it again after six weeks' absence, encased as it had been in a fibreglass cast. I stroked my leg and rubbed it as I lay in bed listening to the rain pouring down and tried to make friends with it again. It was stiff, purple and swollen, scarcely recognizable, and felt oddly detached from me. Recent neuroscientific research has shown that even within a few days of a limb being lost or immobilized the brain starts to re-wire itself, with other areas of the brain taking over the redundant area for the lost or immobilized limb. My slight feeling of estrangement from my leg was almost certainly an aspect of this phenomenon – the phenomenon of 'neuroplasticity', whereby the brain is constantly changing itself.

After a month off, I could now start cycling to work again, proudly displaying my Star Trooper boot to the passing traffic. This first day back at work was a Thursday, my outpatient day, so after the morning meeting I would be in the outpatient clinic.

Once again, the SHOs at the morning meeting were new and I did not recognize any of them. One of them presented the first case.

'There was only one admission last night,' he said, looking at the X-ray screen. 'Not very interesting,' he added. He was sprawled back in his chair, his back turned to us, trying to appear cool but looking instead like an awkward teenager.

'Don't ever say that!' I said. 'Who are you, by the way? And what do you want to be when you grow up?' – a standard question I ask all the new doctors.

'Orthopaedic surgeon,' he told me.

'Sit up straight and look us in the eye when you talk,' I said. I told him that the progress of his medical career was going to depend largely on how well he presented himself and his cases at meetings like this one.

I turned to the registrars, and asked if they agreed and they laughed in polite agreement. I told the chastened SHO to tell us about the patient admitted during the night.

He turned a little sheepishly to face us.

'This is a seventy-two-year-old woman who collapsed at home.' As he spoke he fiddled with the keyboard in front of him and a brain scan started to appear on the wall.

'Hang on!' I said. 'Let's have a little more history before we look at the scan. Do we know her previous medical history, was she fit and independent for her age? In what way did she collapse?'

'Apparently she lived on her own and was self-caring and self-ambulating.'

'Self-catering as well?' I asked. 'And self-cleaning like an oven? Does she wipe her own bottom? Come on, speak English, don't talk like a manager. Are you trying to tell us that she looks after herself and can walk unaided?'

'Yes,' he replied.

'So what happened?'

'Her daughter found her on the floor when she went to visit her. It's not clear how long she had been there.'

'So what's the differential diagnosis for collapse in the elderly?'

The new SHO reeled off a long list of causes and conditions.

'And where was she on the Glasgow Coma Scale?'

'Five.'

'Don't use numbers! They're meaningless. What was she actually doing?'

'No eye opening to pain, no sounds and flexing.'

'That's better,' I said approvingly. 'I can actually see what she was like. Did she have a neurological deficit when you saw her last night on admission?'

He looked embarrassed.

'I didn't look.'

'How did you know her coma scale then?' I asked.

'It was what the doctors at the local hospital said ...' His voice trailed off in embarrassment.

'You should have examined her yourself. But,' I added, feeling the need to wield a carrot after the stick, 'you're here to learn.'

I turned to the registrars who were enjoying the ritual of teaching by teasing of the new SHO.

'Who was on last night?'

David, one of the registrars nearing the end of his six years' training, called out that he had been on call for emergencies.

'She had a right hemiplegia,' he said. 'Her neck was a bit stiff too.'

'What are the other possible signs on examination if she'd had a subarachnoid haemorrhage?'

'They can have subhyaloid haemorrhage in the eyes.'

'Did she?'

'I didn't look. The ward ophthalmoscope was lost ages ago ...'

The woman's brain scan flashed up in front of us.

'Bloody hell!' I said as I looked at it. 'Why on earth did you accept her? It's a massive bleed into the dominant hemisphere, she's seventy-two, she's in coma – we're never going to treat her, are we?'

'Well, Mr Marsh,' David replied a little apologetically, 'the referring hospital said she was sixty-two. She'd been a university lecturer. Pretty smart, the daughter said.'

'Well she's not going to be smart any longer,' my colleague sitting next to me said.

'Anyway,' David said, 'we had some empty beds and the bed managers were trying to put some non-neuro patients in them ...'

I asked if there had been any other admissions.

'The oncologists referred a woman with melanoma,' Tim, one of the other registrars, said, walking to the front of the room to take over from the SHO. He put a brain scan up on the wall in front of us. The scan showed two large and ragged tumours in the brain which were clearly inoperable. Multiple brain tumours are almost always secondary tumours – known as metastases – from cancers that have started elsewhere, such as cancers of the breast or lung or, as in this case, of the skin. Their development signifies the beginning of the end, although in some cases treatment can prolong life by a year or so. 'The referral letter says she drinks 140 units of alcohol a week,' Tim told us. I saw one of the SHOs in the front row doing some rapid mental arithmetic.

'That's two bottles of vodka a day,' she said in slight amazement.

'She had a cerebral met removed at another hospital eighteen months ago,' Tim said. 'And radiotherapy afterwards. The oncologists want it biopsied.'

I asked him what he had told them.

'I said they were inoperable and a biopsy was unnecessary. They are obviously melanoma mets. They might as well make the diagnosis post mortem.'

'I love your positive attitude,' my colleague sitting next to me said. 'So what's the message back to the oncologists?'

'Keep on drinking!' somebody shouted happily from the back of the room.

With no more cases to discuss we filed out of the viewing room to start the day's work.

I stopped at my office to collect my dictation machine.

'Don't forget to remove your tie!' Gail shouted at me through the door to her office.

The new chief executive for the Trust, the seventh since I had become a consultant, was especially keen on the twenty-two-page Trust Dress Code and my colleagues and I had recently been threatened with disciplinary action for wearing ties and wristwatches. There is no evidence that consultants wearing ties and wristwatches contributes to hospital infections, but the chief executive viewed the matter so seriously that he had taken to dressing as a nurse and following us on our ward rounds, refusing to talk to us and instead making copious notes. He did, however, wear his chief executive badge – I suppose just in case somebody asked him to empty a bedpan.

'And your watch!' Gail added with a laugh as I stomped off to see my outpatients.

*

The outpatients wait in a large and windowless room on the ground floor. There are many patients, sitting in obedient silence in rows, as there are many clinics going on all at once in the new, centralized outpatient department. The place has all the charm of an unemployment benefits office although with the added detail of a magazine rack holding leaflets on how to live with Parkinson's disease, prostatism, irritable bowel syndrome, myasthenia gravis, colostomy bags and other unpleasant conditions. There are two large abstract paintings, one purple, one lime green, which were hung on the walls by the hospital arts officer, an enthusiastic woman in black leather trousers, on the occasion of a visit by one of the royals for the official opening of the new building a few years ago.

I made my way past the waiting patients who watched me as I walked to the consulting room. My first sight on entering the consultation room was a Babel-like tower of multicoloured folders containing the patients' notes – a pile rarely less than two feet high – a tower of sheets of paper, bursting out of dog-eared files, in which the recent relevant results have rarely been filed, and if they have been filed, have been filed in such a way that it is usually very difficult to find them. I can learn – usually in entirely random order – about my patient's birth history, and perhaps their gynaecological, dermatological or cardiological conditions, but rarely find information such as when I had operated on the patient, or the analysis of the tumour I removed. I have learnt that it is usually much quicker to ask the patient instead. The Trust has to devote ever-increasing numbers of staff and resources to the constant tracking, searching and transporting of medical notes. The greater part of the notes, I should explain, consist of nursing charts recording the patient's passing of bodily fluids on previous admissions and are no longer of any interest or importance. There must be many tons of such notes being carried around NHS hospitals every day in a strange archival ritual which brings dung beetles to mind, devoted to the history of patients' excretions.

My outpatient clinic is an odd combination of the trivial and the deadly serious. It is here that I see patients weeks or months after I have operated on them, new referrals or long-term follow-ups. They are wearing their own clothes and I meet them as equals. They are not yet in-patients who have to submit to the depersonalizing rituals of being admitted to the hospital, to be tagged like captive birds or criminals and to be put into bed like children in hospital gowns. I refuse to have anybody else in the room – no students, no junior doctors or nurses – only the patients and their families. Many of the patients have slowly growing brain tumours, too deep in the brain to operate upon and yet not growing fast enough to justify the palliative cancer treatment of radiotherapy or chemotherapy. They come to see me once a year for a follow-up scan to see if their tumour has changed or not. I know that they will be sitting outside the consultation room, in the dark and dreary waiting area, sick with anxiety, waiting to hear my verdict. Sometimes I can reassure them that nothing has changed, sometimes the scan shows that the tumour has grown. They are being stalked by death and I am trying to hide, or at least disguise, the dark figure that is slowly approaching them. I have to choose my words very carefully.

Since neurosurgery deals with diseases of the spine as well as of the brain, part of any neurosurgical outpatient clinic is spent talking to patients with back problems, only a few of whom need surgery. With one patient with a brain tumour I will be trying to explain that his or her life is probably coming to an end, or that they require terrifying surgery to their brain, whereas with the next I will be telling them, struggling to appear sympathetic and uncritical, that their backache is perhaps not as terrible a problem as they feel it to be and that life can perhaps be worth living despite it. Some of the conversations I have in the clinic are joyful and some absurd and others can be heart-breaking. They are never boring.

Having looked at the pile of notes with a feeling of slight despair, I sat down and turned on the computer. I returned to the reception desk to look at the list of patients for the clinic and find out who had arrived. All I could see were several blank sheets of paper. I asked the receptionist for my clinic list. He looked a little embarrassed as he turned over one of the blank sheets of paper to reveal another sheet beneath with a list of the patients coming to see me.

'Corporate outpatient management has said we must keep the patient's names covered to preserve confidentiality,' he said. 'There's some target for it. We've been told to do it.'

I called out the name of the first patient in a loud voice, looking round at the assembled patients waiting to see me. A young man and an elderly couple hurriedly got up from their chairs, anxious and deferential in the way we all are when we go to see doctors.

'How's that for confidentiality?' I muttered to the hapless receptionist. 'Maybe the patients should be identified only by numbers like in a VD clinic?'

I turned away from the reception desk.

'I'm Henry Marsh,' I said to the young man as he came up to me, becoming a kind and polite surgeon, instead of an impotent and angry victim of government targets. 'Please follow me.' We walked round to the office, his elderly parents following us.

He was a young policeman who had suffered an epileptic fit several weeks ago – suddenly, out of the blue, changing his life forever. He was taken to his local A&E department where a brain scan showed a tumour. He had recovered from the fit, and as the tumour was a small one, he was sent home and a referral made to the regional neurosurgical centre. It took a while for the referral letter to be passed on to me so he had to wait two weeks before we finally met – two weeks waiting to hear, in effect, whether he would live or die since none of his local doctors would have known enough about brain tumours to be able interpret the scan with any confidence.

'Do please sit down,' I said, indicating the three chairs in front of my desk, with its tower of notes and slow computer.

I went over the history of the fit briefly with him and his family. As is usually the case with epilepsy it had been more frightening for his mother, who had witnessed the fit, than for the man himself.

'I thought he was going to die,' she said. 'He stopped breathing and his face went blue though he was better by the time the ambulance came.'

'I just remember waking up in the hospital. I then had the scan,' the young policeman said. 'I've been fearing the worst since then.' His face was desperate with the hope that I could save him, and the fear that I might not.

'Let's look at the scan,' I said. I had seen it two days earlier but I see so many scans every day that I have to have them immediately in front of me whenever I see a patient if I am not to make a mistake.

'This may take a while,' I added. 'The scans are on the computer network of your local hospital and this is then linked over the net to our system ...' As I spoke I typed on the keyboard looking for the icon for his hospital's X-ray network. I found it and summoned up a password box. I have lost count of the number of different passwords I now need to get my work done every day. I spent five minutes failing to get into the system. I was painfully aware of the anxious man and his family watching my every move, waiting to hear if I would be reading him his death sentence or not.

'It was so much easier in the past,' I sighed, pointing at the redundant light-screen in front of my desk. 'Just thirty seconds to put an X-ray film up onto the X-ray screen. I've tried every bloody password I know.' I could have added that the previous week I had had to send four of the twelve patients home from the clinic without having been able to see their scans, so that the appointments had been entirely wasted and the patients made even more anxious and unhappy.

'It's just like this with the police force,' the patient said. 'Everything's computerized and we are constantly being told what to do but nothing works as well as it used to …'

I rang Gail but she was unable to solve the problem. She gave me the number of the X-ray department but when I tried it I only got an answering machine in reply.

'Excuse me,' I said. 'I'll go upstairs to see if I can get one of the X-ray secretaries to help.' So I hurried past the waiting patients in the subterranean waiting room and ran up the two flights of stairs to the X-ray department – it is quicker than going by the lifts and without a condescending voice telling me to wash my hands.

'Where is Caroline?' I shouted as I arrived at the X-ray reception desk, a little out of breath.

'Well, she's about somewhere,' came the reply so I headed off round the department and eventually I found her and explained the problem.

'Have you tried your password?'

'Yes, I bloody well have.'

'Well, try Mr Johnston's. That usually works. Fuck Off 45. He hates computers.'

'Why forty-five?'

'It's the forty-fifth month since we signed onto that

hospital's system and one has to change the password every month,' Caroline replied.

So I ran down the corridor and down the stairs and past the waiting patients back to the consultation room.

'Apparently the best password is Fuck Off 45,' I told the patient and his parents, who were still waiting to hear his possible death sentence. They laughed nervously.

I duly typed in 'Fuck Off 45' but, having thought about it, and having told me that it was 'checking my credentials' the computer told me that the password was not recognized. I tried typing in Fuck Off 45 in many different ways, upper case, lower case, with spaces, without spaces. I typed in Fuck Off 44 and Fuck Off 46 but without success. I ran back upstairs a second time, followed by the curious, anxious eyes of the patients in the waiting area. The clinic was now running late and the number of patients waiting to see me was steadily growing.

I went back to the X-ray Department and found Caroline at her desk. I told her that Fuck Off 45 did not work.

'Well,' she sighed, 'I'd better come and look. Maybe you don't know how to spell Fuck Off.'

We went downstairs together and returned to the consultation room.

'Now that I think of it,' she said. 'It might have become Fuck Off 47.' She typed in 'Fuck Off 47' and the computer, having checked my credentials – although they were really Mr Johnston's – to its satisfaction, finally downloaded the menu for the X-ray department at the patient's hospital.

'Sorry about that!' Caroline said with a laugh as she left the room.

'I should have thought of that myself,' I said, feeling very stupid, as I downloaded the patient's brain scans.

It may have taken a long time to get the brain scans onto the computer screen but it did not take long to interpret them. The patient's scan showed an abnormal area – a little like a small, white ball, pressing on the left side of his brain.

'Well,' I said, knowing what had been preying on his mind for the last two weeks, and more particularly over the past fifty minutes, 'it doesn't look like cancer ... I think everything's going to be OK.' All three of them sat back a little in their chairs as I said this, and the mother reached out for her son's hand, and they smiled to each other. I felt considerable relief myself. I often have to reduce people to tears as they sit opposite me in the outpatient room.

I explained to them that the tumour was almost certainly benign and that he would need an operation to remove it. I added a little apologetically that the operation had some serious risks. In a reassuring tone of voice I explained that the risk of leaving him paralysed down the right side of the body – as though he'd had a stroke – and maybe unable to speak – was 'not more than five per cent'. It would sound very different if instead I said 'as much as five per cent' in a suitably grim tone.

'All operations have risks,' said his father, as almost everybody does at this stage of the discussion.

I agreed, but pointed out that some risks are more serious than others and the trouble with brain surgery is that even if little things go wrong the consequences can be catastrophic. If the operation goes wrong it's a one hundred per cent disaster rate for the patient but still only five per cent for me.

They nodded mutely. I went on to say that the risks of the operation were very much smaller than the risks of doing nothing and letting the tumour get bigger – eventually even benign tumours can prove fatal if they grow large enough as the skull is a sealed box and there is only a limited amount of space in the head.

We talked a little more about the practicalities of the operation and I then took them round to Gail's office.

The next patient was a single mother with back pain who had undergone two ill-advised operations on her back in the private sector. There is a well-recognized syndrome called 'the failed back syndrome' which refers to people with backache who have undergone spinal surgery which has not worked (and which in many cases seems to have made their pain worse).

She was thin and had the haunted expression of somebody in constant pain and deep despair. I learned a long time ago in the outpatient clinic to make no distinction – as some condescending doctors still do – between 'real' or 'psychological' pain. All pain is produced in the brain, and the only way pain can vary, other than in its intensity, is how it is best treated, or more particularly in my clinic, whether surgery might help or not. I suspect that many of the patients in my clinic would be best treated by some form of psychological treatment but it is not something I am in a position to provide in a busy surgical outpatient clinic, although I will often find myself having to spend longer talking to the patients with back pain than those with brain tumours.

She started to cry as she spoke.

'My pain is worse than ever, doctor,' she said, her elderly mother who was sitting beside her nodding anxiously as she spoke. 'I can't go on like this.'

I asked her the usual questions about her pain – a list one learns early on as a medical student – questions as to when she got the pain, whether it went down her legs, what type of pain it was and so on. With experience one can often predict the answers just by looking at the patient and as soon as I had seen her tearful, angry face as she dramatically limped down the corridor behind me to the consultation room I knew that I was not going to be able to help her. I looked at the scan of her spine which showed plenty of space for the nerves, but also the excavations and crude metal scaffolding carried out by my surgical colleague elsewhere.

I told her that if an operation fails to work there are two diametrically opposite conclusions to be drawn – one is that it wasn't done well enough and needs to be done again, the other is that surgery wasn't going to work in the first place. I told her that I didn't think that another operation would help her.

'But I can't go on like this' she said angrily. 'I can't do the shopping, can't look after the kids.' The tears started to stream down her face.

'I have to do that,' her mother said.

With patients like this, when I know that I cannot help them all I can do is sit quietly, trying to stop my eyes drifting away out of the window, over the car park, over the hospital's perimeter road, towards the cemetery on the other side, as the patients pour out their misery to me, and wait for them to finish. I then have to find some form of words with some expression of sympathy to bring the hopeless conversation to a close and suggest that their GP refer them to the Pain Clinic with little hope that their pain can be cured.

'There is nothing dangerous about the situation in your back,' I will say, taking care not to say that the scan is essentially normal, which it often is. I will deliver a little speech on the benefits of exercise and, in many cases, losing weight, but this advice is rarely well-received. I do not pass judgement on these unhappy people – as I did when I was younger – but instead I feel a sense of failure and also occasionally of disapproval for surgeons who have operated on patients like this, especially when it has been done – as is often the case – for money, in the private sector.

The next patient was a woman in her fifties who had undergone removal of a large benign brain tumour twenty years earlier by a colleague who had retired many years ago. Her life had been saved but she had been left with chronic facial pain. Every possible form of treatment had failed. The pain had developed because the sensory nerve for one side of her face had been severed as the tumour had been removed – an occasionally unavoidable problem for which surgeons use the word 'sacrifice'. This leaves the patient with severe numbness on that side of the face, an unpleasant phenomenon, though one with which most people come to terms. A few, however, do not and instead they are driven almost mad by the numbness, the Latin medical name for this, *anaesthesia dolorosa* – painful loss of feeling – expressing the paradoxical nature of the problem.

This patient also spoke at interminable length – describing the many unsuccessful treatments and drugs she had had over the years, and the uselessness of doctors.

'You need to cut the nerve doctor,' she said. 'I can't go on like this.'

I tried to explain that the problem had arisen precisely because the nerve had been cut, and told her about phantom limb pain, where amputees experience severe pain in an arm or leg which no longer exists as a limb in the outside world but which still exists as a pattern of nerve impulses in the brain. I tried to explain to her that the pain was in her *brain* and not in her face but the explanation was lost on her, and judging from her expression she probably thought I was dismissing her pain as being *'all in the mind'*. She left the room as angry and dissatisfied as she had been when she had arrived.

One of several patients with brain tumours whom I saw for regular follow-up was Philip, a man in his forties with a tumour called an oligodendroglioma whom I had operated upon twelve years earlier. I had removed most of the tumour but it was now growing back. He had recently had chemotherapy for this, which can slow down the rate of recurrence, but both of us knew that the tumour would eventually kill him. We had discussed this on previous occasions and there was little to be gained by going over it again. Since I had been looking after him for so many years we had got to know each other quite well.

'How's your wife?' was the first thing he said to me as he came into the room and I remembered that when we had last met a year earlier I had been phoned by the police in the middle of the conversation to be told that my new wife Kate, whom I had met a year after my first marriage ended, had just been admitted to hospital with an epileptic fit.

'Nothing to worry about,' the policeman had said, trying to be helpful. I had brought the consultation with Philip to a rapid conclusion and rushed off to the A&E department in my hospital where I found Kate looking quite unrecognisable with her face covered in dried blood. She had had an epileptic fit in the Wimbledon shopping centre and bitten straight through her lower lip. Fortunately she had not suffered any major harm and one of my plastic surgical colleagues came and stitched up the laceration. I arranged for her to see one of my neurological colleagues.

It was a difficult time – many brain tumours first declare themselves with epileptic fits as I knew all too well and I also knew well enough from my experience with my son that my being a doctor did not make me or my family immune to my patients' diseases. I did not share these thoughts with Kate and told her that the scan was just a formality, hoping to spare her some anxiety. Kate is an anthropologist and bestselling writer, with no medical background, but I had underestimated her powers of observation. She told me later that she had picked up enough neurosurgical knowledge to know that brain tumours often 'present with' epilepsy. We had to wait a week before she had a brain scan, during which we carefully hid our fears from each other. The scan was normal: there was no tumour. The thought that so many of my patients have to go through the same hell of waiting for the results of scans as Kate and I did is not a comfortable one – and most of them must wait much longer than a week.

Touched by Philip's remembering this, I told him that she was fine and that her epilepsy was currently under control. He told me that he was continuing to have minor fits several times a week, and that his business had gone bankrupt because he had lost his driving licence.

'Lost a lot of weight with the chemo though,' he said with a laugh. 'I look a lot better, don't I? Made me pretty sick. I'm alive. I'm happy to be alive. That's all that matters, but I need to get my driving licence back. I get only sixty-five pounds a week in benefits. It's not exactly easy to live on that.'

I agreed to ask his GP to refer him to an epilepsy specialist. Not for the first time, I thought of the trivial nature of any problems that I might have compared to my patients' and felt ashamed and disappointed that I still worry about them nevertheless. You might expect that seeing so much pain and suffering might help you keep your own difficulties in perspective but, alas, it does not.

The last patient was a woman in her thirties with severe trigeminal neuralgia. I had operated on her the previous year and vaguely remembered that she had then come back a few months later with recurrent pain – the operation occasionally fails – but I could not remember what had happened afterwards. I fumbled through the notes unable to find anything of help. I prepared a speech of apology expecting her to look miserable with pain and disappointment. Now, however, she was quite different. I expressed surprise at how well she looked.

'I've been absolutely fine since the op,' she declared.

'But I thought the pain had come back!' I said.

'But you operated again!'

'Did I really? Oh I am sorry – I see so many patients that I tend to forget ...' I replied.

I pulled her notes off the pile and spent several minutes failing to find something about her having had a second operation. Out of the inches of paper a brown tab stuck out – one of the few documents that the Trust has designed that is easily located.

'Ah!' I said, 'Look. I may not be able to find the operating note but I can tell you that you passed a type-4 turd on 23 April ...' I showed her the elaborate hospital Stool Chart, coloured a sombre and appropriate brown, each sheet with a graphically illustrated guide to the seven different types of turd in accordance with the classification of faeces developed, according to the chart, by a certain Dr Heaton of Bristol.

She looked at the document with disbelief and burst out laughing.

I pointed out to her that she had passed a type-5 next day – 'small and lumpy, like nuts' according to Dr Heaton – and showed her the accompanying picture. I told her that, as a brain surgeon, I couldn't give a shit about her bowel movements although the Trust management clearly considered it a matter of deep importance.

We laughed together for a long time. When we had first met, her eyes were dull with pain-killing drugs and if she tried to talk her face would contort with agonizing pain. I thought how radiantly beautiful she now looked. She stood up to leave and went to the door but then came back and kissed me.

'I hope I never see you again,' she said.

'I quite understand,' I replied.

CODA

I had first met Will two years earlier when he had been referred to me by my radiotherapy colleague. He was, I suppose, what you might call a gentleman plumber – at least, he did not fit my stereotype of what a plumber should look like. My height, quite slim, with a thoughtful expression and an educated accent, he was in his forties, with a younger wife and two small children. It was only after the operation that I learned he had once studied woodcut engraving in Japan. Perhaps because we were fellow craftsmen we hit it off as soon as we met. Besides, his quiet and stoical manner was exactly what doctors like to find in their patients.

He had originally undergone surgery at another major neurosurgical hospital for his tumour – a petro-clival meningioma, a benign tumour growing underneath the brain and pressing upwards onto it. I did not know the details when I first met him, but Will told me that he had almost died from blood loss during the operation and the surgeon had had to leave much of the tumour behind. The tumour had 'presented', as doctors put it, with progressive deafness in his right ear, and after the operation the right side of his face was very numb and he had double vision. This all slowly got better, but he remained deaf in the right ear. Since most of the tumour was still in place he was sent for radiation treatment, and when it became apparent that this was not working – the tumour was continuing to grow – he was sent to see me.

Petro-clival tumours are among the most difficult operations in neurosurgery. Technological progress has rendered many of the most exquisitely difficult neurosurgical procedures – such as clipping aneurysms of the basilar artery, the major artery that keeps us alive and conscious – almost obsolete. Although neurosurgical operations are probably less dangerous than in the past, there are exceptions – such as Will's tumour. The problem with such tumours is that they grow in a corner of the skull called the cerebellopontine angle, known as the CPA by neurosurgeons. This is a small space a few millimetres across, deep inside on either side of the head, just behind each ear. It is traversed by the nerves for moving and feeling in the face, for hearing, and one of the nerves for moving the eyes, as they emerge from the brain to enter tunnels in the skull on their way to the body outside. The anatomical names for these cranial nerves are the facial, trigeminal, acoustic and trochlear nerves. They are very small and delicate structures, two to three millimetres in size, and very easily damaged. If damage occurs half the face and tongue will be numb and paralysed, the patient will have disabling double vision, and will be left deaf in one ear.

These are not easy problems with which to live and are life-changing events, much for the worse. But there is an even greater risk with a CPA tumour like Will's: if it is large enough – as his was – it will press upon the basilar artery, as it runs up in the middle of the head to take blood to the brainstem, the part of the brain that contains all the control mechanisms for consciousness, and breathing and blood pressure and the beating of the heart. Damage to the basilar artery and its many small branches, known as perforators, will usually lead to death or a terrible stroke, sometimes of the 'locked-in' variety, where the victim cannot move at all, and yet is fully conscious. This is what happened to the French writer Jean-Dominique Bauby, who suffered a spontaneous stroke in the basilar artery, and described this awful condition in the book *The Diving Bell and the Butterfly* shortly before he died. He wrote it by blinking an eyelid, the only movement he could make, blinking for yes or no as he was shown letters on a board. It is what had happened to the schoolteacher on whom I had operated many years earlier and who had been left in a persistent vegetative state – his tumour had been very similar to Will's.

Re-operating on a tumour is almost always more difficult than operating for the first time because scar tissue will destroy what surgeons call the tissue planes – the normal boundaries between the tumour and the adjacent brain and cranial nerves. So although I knew that Will would slowly die if his tumour was not removed, I did not want to do the operation – he was bound to be left worse off to some extent, however well the operation went. He was still well, deaf in one ear, but working. The brain scans showed that the tumour was only growing very slowly, though this inexorable growth was certainly a death sentence. I was soon to retire and I longed not to have to go through the misery of wrecking another patient before I finally hung up my gloves and stopped operating.

'There is no great hurry to operate,' I told him when I first saw him and his wife. 'If we postpone the operation until the tumour is only a little bit larger it won't make a big difference to the risks. We can just get scans every six months or so and postpone a difficult decision. But I'm afraid it's not a question of whether to operate, it's a question of when.'

He took this all in quietly but his wife looked terrified. I have usually found with married couples that one will be frightened and the other will appear calm – though which way around it will be is unpredictable.

'And the risks?' he asked.

'Well, blood loss, as you already know, but the main question is whether the tumour is soft or not and whether it is stuck to the brainstem and cranial nerves. You can't judge that from the scan.'

'So you can't tell until you get in?'

'Exactly. If the tumour's hard or stuck to the brainstem there's a real danger of death or a major stroke.'

'But I've heard that you're retiring soon,' he said, 'and I want you to do the operation.'

'But I'm not the best,' I replied a little defensively. 'Though with tumours like this nobody is.'

As I said this I thought of the presentations I had heard over the years at international conferences by the great names of neurosurgery and the breathtaking results they claimed with CPA tumours like Will's – although rarely, if ever, with a word about any bad results they might have had along the way. At such conferences surgeons will gather afterwards in the bar and swap stories of how they have come across the great surgeons' great disasters. But I know there is always an element of jealousy and sour grapes to such conversations as we try to reassure ourselves that maybe the great surgeons are not *that* much better than we ordinary surgeons. Will's tumour was indeed very similar to the one that I had operated upon with a disastrous result twenty years earlier, though I had had some relatively better results since then. Despite my misgivings, it was clear that I was now stuck with Will and his tumour. I liked him greatly, and he certainly trusted me. It was difficult to know whether this made things better or worse.

'I've got a lot of work on at the moment,' he said. 'Perhaps I could see you again in six months' time with a scan?'

So for the next one and a half years I would see him every six months with a new scan. Each time the tumour would be

just a little larger, and each time he was calm and matter-of-
fact and his wife looked terrified. A few months before I was
to retire we agreed that I should operate.

I had slept very well and had gone to work in a surprisingly peaceful mood – surprisingly since I knew the operation
would be terrible. Shortly after first meeting Will I had written to the surgeon who had originally operated on him asking
him about what had happened. After many months came the
reply that the tumour had indeed been 'very vascular' – that
is, it had bled a lot – and that Will had almost 'exsanguin-
ated' – that is, bled to the death on the table.

At least I was forewarned and there were various precau-
tions I could take that might give me an edge. I would do
the operation with Will in the sitting position. If the patient
is sitting upright during the procedure the tumour usually
bleeds less and also the blood runs downwards so that you
can see what you are doing. If the patient is lying flat – which
is how most surgeons do this kind of surgery – their head
fills with blood and there is a constant struggle to keep the
field clear with a sucker so that you can safely identify the
tumour and the surrounding critical nerves and arteries. The
sitting position, on the other hand, is not without its own
risks which is why few surgeons use it. I would also get the
tumour embolized, which is done by injecting glue or microscopic particles into the blood vessels taking blood to the
tumour. They act like miniature corks in a bottle's neck to
block off the flow of blood to the tumour. The procedure is
done by radiologists under X-ray control. It is a little like a
video game where they manipulate an extremely small catheter into the blood vessels of the patient's brain, watching
the tip of the catheter on a video X-ray screen as they twirl
and push the end of the catheter between their fingers at the
point of its insertion into the patient's groin, thus controlling

the tip three feet away in the patient's brain. It is as skilful as surgery and also not without risk – it is possible inadvertently to block off the arteries that take blood to the brain rather than to the tumour and cause a stroke.

I had planned all this days in advance with my radiological colleagues so, once Will was anaesthetized, he was taken to the X-ray department and Jeremy, one of my neuroradiology colleagues, inserted a catheter into Will's femoral artery and all the way up into the blood vessels for his brain. This had been started in the X-ray suite while I was operating in the main theatre. Once I was finished – it was a straightforward operation for trigeminal neuralgia – I walked the short distance down the corridor to the Neuro-X-ray department. I found Jeremy in the control room looking at a bank of computer screens showing the black-and-white images of the blood vessels of Will's brain. The brain's blood vessels – or vasculature in medical language – are usually structured like a tree, with trunks and branches gradually narrowing into smaller and smaller twigs. In the centre of Will's brain, however, a great black tangle of arteries – black from the X-ray dye injected through the catheter – hung like a large and ominous bunch of mistletoe: these were the blood vessels in the tumour.

'Bloody hell!' I said to Jeremy. 'That's one hell of a blush' – blush being the word we use for the way that a tumour will blush with dye on angiography. 'But we knew it was vascular. Can you embolize it?'

''Fraid not,' Jeremy said, 'the vessels are all wrong. The particles would fly off all over the place. Cause a basilar stroke.'

'Damn! I'm not looking forward to this, you know.'

At that point one of the other radiologists came into the control room. He leant forward and peered at the screens.

'We could try fibre coils,' he said. 'They'll last a short time but if you take him straight to theatre it might help.'

'Well, do your best,' I said, and I went off to see if my other patient had woken up and then went down to my office to clear some paperwork.

Two hours later I was back in the theatre. Will, anaesthetized, with any number of tubes, lines and wires tied and taped and even stitched to his unconscious and naked body, was on a trolley, waiting to be lifted over onto the operating table.

Judith, the anaesthetist, was standing beside the anaesthetic machine adjusting various controls.

'Big drips?' I asked her, which was a way of asking whether she had inserted unusually large intravenous catheters to allow for rapid and massive blood transfusion if necessary. 'Lots of blood cross-matched?

'Yes,' she said, rolling her eyes up ever so slightly.

'Did Jeremy and Andy manage to embolize it at all?' I asked.

'Well, they weren't sure.'

'Oh,' I replied.

With the help of my assistant and the theatre staff we got Will onto the table and sat him upright, always quite a complicated process. The operation started with a near disaster. I had assumed my colleague at the other neurosurgical hospital would have done the same kind of craniectomy – the opening of the skull – as I do, but it was subtly different and perhaps the scar of the incision had shifted over the underlying skull. I became completely lost in the muscles of his neck – muscles you must work through to get at the bottom of the skull when approaching the CPA. I came within a hair's width of severing the vertebral artery, one of the major tributaries in the neck for the all-important basilar artery, before I found my bearings. Surgical experience is all about familiarity and recognition and it is alarmingly easy, when re-operating, to get lost, especially if you have not done the original operation yourself. I had been so worried about the tumour itself and the risk of haemorrhage that I had not given sufficient thought to the beginning of the operation and the approach. But things improved and I soon found the tumour easily enough, deep within Will's head, although I could not identify any of the cranial nerves – they were all buried in scar tissue.

'The nerves will be hidden somewhere at the bottom,' I said to Haru, the registrar who was assisting me. 'We'll start dissecting out the tumour higher up and hope to find a plane and work downwards, finding the nerves.'

With operations like this you are operating more or less at the limit of the length of your instruments, using a microscope – seven or eight inches down, through a gap between the brain and the inside of the skull of only a few millimetres. The tumour was in the right CPA, so on the left was the part of the brain called the brainstem, which was distorted and indented by the tumour. The brainstem keeps the rest of the tree alive and awake. A small area of damage to it is almost always disastrous. The blood supply to the brainstem, which keeps the brainstem alive, is provided by the basilar artery.

I started by debulking the tumour, taking out its centre so as to collapse it away from the brain and the important nerves that I could not find, and then gently teasing the tumour's capsule – in effect its skin – away from where it was indenting the brainstem. With a pair of microscopic tumour forceps I grasped the capsule and pulled it away from the brain, hoping that the brain would not tear in the process. Initially the plane between the brainstem and capsule seemed rather good, and the tumour capsule separated away cleanly from the brainstem. My morale rose, I began to allow myself the hope that the operation was going to be a success after

284

all. But, as is so often the case with these tumours, things soon became steadily worse and the side of the brainstem looked increasingly bruised and battered, nor could I see any of the cranial nerves which I was so anxious not to hurt. Ominously, I could see a few broken strands of what looked like nerve stuck to the side of the tumour. Despite this I could not help but experience the intense excitement, concentration and fierce joy I always feel when I do these dangerous operations.

'One's on a higher plane of existence when operating like this,' I said to Haru, only half-joking, turning aside for a moment from the microscope's eyepieces. 'It's utterly addictive.' Haru said nothing but I knew that he understood me perfectly.

I continued to remove the tumour, bit by bit, over the next four hours. Eventually I had removed it all, apart perhaps from a small piece buried in the skull which would be most unlikely to cause any problems during the rest of Will's life. With the tumour now all gone there was a spectacular view of the basilar artery with its little perforating branches going to the brainstem.

'Have a look at that, Haru,' I said happily. 'That's not a sight you'll often get to see. I'm afraid we have to assume that the cranial nerves have all gone – so he'll have a full house of nerve palsies: a completely dead right side to his face, no movement on that side, and double vision.'

'But you've saved his life,' Haru said.

'Well, yes, but a numb, paralysed face is a life-changing event. But you're right – and he accepts that. At least I think he does. Judith,' I said to the anaesthetist, 'it's all out. Can you bring the blood pressure up for haemostasis?'

Haemostasis is the vital bit at the end of the operation when we make sure that all the bleeding has stopped. Any bleeding after the operation is a life-threatening emergency and if it happens we must rush the patient back to the operating theatre. We take great care, therefore, at the end of the operation to make sure that all the bleeding has stopped.

I leaned back in my operating chair, my arms on the armrests and waited for the blood pressure to come back to a normal level. I regretted the damage to his face but I didn't see how I could have avoided it and I did not feel too bad about it. While I thought these thoughts I noticed to my annoyance that bright red blood was starting to trickle out of the bottom of the wound. Probably just the muscle bleeding, I told myself, although I was surprised at the way in which it was so bright. But when I brought the microscope back in I found to my dismay a fine spray of blood filling the tumour cavity, brilliant red droplets of arterial blood looking disconcertingly pretty as they danced and darted about, glittering and magnified in the microscope's light. I traced the origin of this dance down to the basilar artery.

'Oh!' I said to Haru. 'It's coming from the basilar. How the hell did that happen? I didn't even touch it.'

There was a hole in Will's basilar artery, pumping out a fine jet of blood that scattered into a dance of death, a balletic cloud of little blood droplets – a death sentence, a catastrophic stroke at best.

'Judith,' I said quietly, 'I've got bleeding from the basilar.'

'OK,' she said, but I could hear the disappointment in her voice.

I don't know how it happened – perhaps a perforator I had unknowingly severed earlier which had opened up as Judith brought up the blood pressure for haemostasis. It was highly unlikely that the bleeding two-millimetre stump, where somehow I had managed to tear the vessel off the great trunk of the basilar, did not play some vital role in keeping Will alive and well.

'You mustn't coagulate arterial bleeding like this,' I said to

Haru, as I struggled to get my instruments up to the basilar through the tiny space left by removing the tumour. It felt as though it were miles away, not a mere six inches – they could only just reach.

'If the bleeding is coming from very close to the artery like this, diathermy will just burn a bigger hole and the bleeding will get worse,' I explained.

'Vivienne,' I said to the scrub nurse, 'Surgicel, then a very small piece of muslin.'

I do not have time to feel nervous or frightened when something like this happens and my actions feel almost like reflexes, and yet, if I think about it, they are reflexes based on almost four decades of operating and many mistakes from which I have learned many bitter lessons. I waited a little impatiently while Vivienne prepared a microscopic piece of muslin – simple cloth, very old fashioned, but very effective for stopping arterial bleeding. While I waited I watched the spray of blood dancing in the microscope's light, already resigning myself to the fact that a disaster had occurred, that there were weeks of grief and misery to come. Although it felt like hours, it was probably only a matter of seconds before Vivienne handed me a pair of microscopic forceps with the muslin and I was able to stop the bleeding.

'Gentle packing and time,' I said to Haru.

We waited a good ten minutes before deciding that there was no more bleeding and I completed the operation in a sad and sombre state of mind. There was none of the usual cheerful chatter as Haru and I closed the wound. I could not see any way in which Will would not either die or be left dreadfully disabled. As we feared the worst Judith decided to wake him up slowly on the ITU instead of reversing the anaesthetic while Will was still in the operating theatre. It would be several hours before I could know whether he would live, or whether he had suffered a major stroke.

I found his wife in the brightly lit corridor outside, sitting in a line of otherwise empty chairs – it was already seven o'clock in the evening. She watched me walk slowly towards her. It was hard to know what facial expression I should choose – I could not very well smile happily. I tried to look neutral but I don't doubt she immediately knew that all was not well.

'I've removed it,' I said. 'But it was very difficult. The bleeding wasn't too bad but everything else was difficult. I just don't know yet how he will do. He'll certainly have half his face paralysed and numb. It's too early to tell if he's suffered a big stroke or not, but I'm afraid he might have.'

She thanked me, over and over again, I suppose unconsciously hoping that this would save her husband. I threw my arms around her and we hugged. I think she understood that I was almost as frightened and worried as she was.

Back in the operating theatre I sat by the reception desk and various juniors stopped by to chat. They asked me how the operation had gone.

'There was a bleed from the basilar,' I said. 'That usually means death or a stroke.'

'That's tough,' said one of juniors, adding with an apologetic laugh, 'and here I am feeling sorrier for the surgeon than the patient.'

My juniors looked sympathetically at me.

'No, it comes with the territory,' I said. 'It's brain surgery.'

I felt a sudden surge of affection for my team of my juniors and their concern for me.

'You know what I'll miss most when I retire next year?' I said. 'It's you lot. Feeling part of a family, a band of brothers – and sisters. Though I admit I'm the big silver-backed gorilla, and you're here to look after me.'

*

After an hour I went round to the ITU. To my amazement, Will seemed to be waking up without a major stroke. He was still unconscious, but if I squeezed his limbs so as to inflict a little pain he withdrew them purposefully, which he would not be able to do if he had suffered a major stroke. I knew he would at least have a facial paralysis and probably many other problems as well, but I cycled home in the dark feeling much less unhappy.

When I returned to the hospital at 10 p.m. I found Will awake, talking a few words and moving his limbs though looking and obviously feeling awful, vomiting into a grey cardboard vomit bowl which his nurse held in front of him. I thought how a facial paralysis was a reasonable price to pay for his life. I gave him a thumbs-up – he was scarcely up to saying much – and left. As I left the hospital, however, I realized that I had forgotten to instruct the ITU nurse about caring for his numb and paralysed eye so I went back to the ITU. I told the nurse what to do – she looked a little puzzled – and when I asked Will to move his face I saw that there was not a trace of weakness. Without ever having seen it I had managed to avoid cutting the nerve for the muscles of his face.

And although I had slept so well the night before the operation I had a restless night after it, not because I was overjoyed or even just relieved – my usual feeling now if a big case goes well – but because I was confused, disconcerted. I kept on waking and repeating to myself over and over again, 'I can't believe it. I can't believe it'. I had prepared myself for the torture of seeing him on my daily ward round, damaged and suffering, for weeks on end and now it would not be necessary.

When I saw him the next morning he was already sitting up in bed, eating some cornflakes, looking a bit bleary-eyed.

He squinted at me as he spoke, closing one eye, since he had double vision.

'You're a lot better than I expected,' I told him, 'but you'll feel awful for many weeks, and probably very unsteady on your legs.'

He rubbed the right side of his face with his hand – it was clear that it was completely numb. 'I'm afraid I don't think the numbness will ever get better,' I said. 'The double vision might. If not the eye doctors can possibly help.'

'Well,' he said, his words a little slurred since the right side of his tongue was as numb as his face. 'I'm alive. Thank you.'

ACKNOWLEDGEMENTS

I hope that my patients and colleagues will forgive me for writing this book. Although the stories I have told are all true I have changed many of the details to preserve confidentiality when necessary. When we are ill our suffering is our own and our family's, but for the doctors caring for us it is only one among many similar stories.

I have been greatly helped by my wise agent Julian Alexander and my excellent editor Bea Hemming. How much worse the book would have been without their guidance! Several friends kindly read drafts of the book and made invaluable suggestions, in particular, Erica Wagner, Paula Milne, Roman Zoltowski and my brother Laurence Marsh. Geoffrey Smith was not only responsible for a beautifully made and very successful film about Igor and me, *The English Surgeon*, but also played an important role in the evolution of the book. Throughout the twenty-seven years of my time as a consultant neurosurgeon I have had the good fortune to have Gail Thompson as my secretary, whose support, efficiency and care for patients has been second to none.

The book would never have been written without the love, advice and encouragement of my wife Kate, who also came up with the title, and to whom it is dedicated.